PI AND PHI

A COY ROMANCE OF
TWO NUMBERS

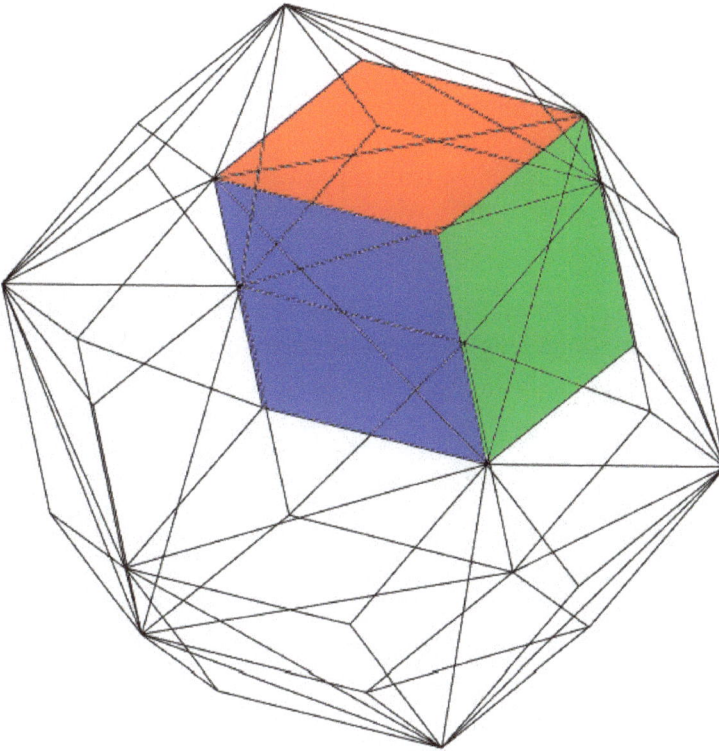

JAMES R WARREN

BLOXWICH
2023

First Published in the United Kingdom in 2023 by Midland Tutorial Productions

Second Edition 1 September 2023

File Prefix Code: PIPHI2

ISBN 978 1 915750 06 8

Midland Tutorial Productions Publishers
31 Victoria Avenue
Bloxwich
Walsall
WS3 3HS
United Kingdom

M┃DLAND
┼UTORIAL

PI AND PHI

Second Edition

James R Warren

MIDLAND TUTORIAL PRODUCTIONS
BLOXWICH

Other Books By James R Warren

Boscawen-Ûn (First and Second Editions)
Beyond Tourist Britain
Gleanings as I Pass
Exordium
Meditations
Gamma Solution
Moddeshall Hydropower
Unreasonable Mathematics
Mathematical Explorations
Researches: Volume One
Researches: Volume Two
Researches: Volume Three
Researches: Volume Four

To The Glory of The Loving God

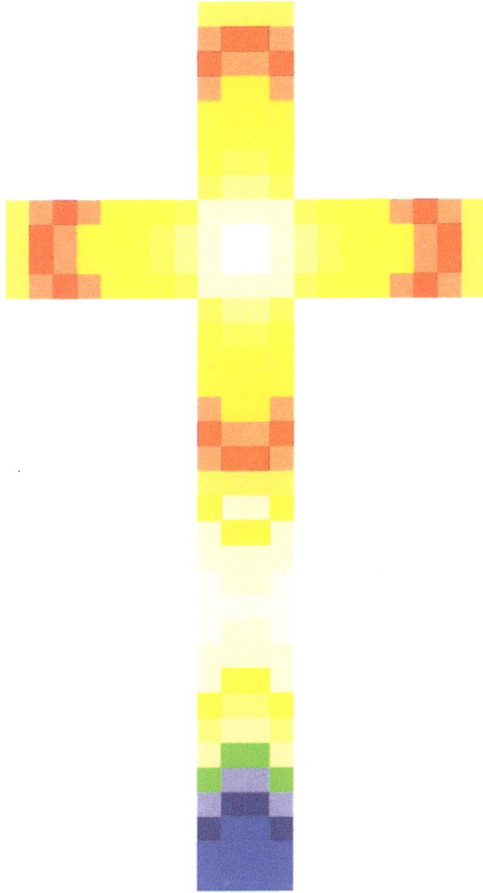

Who Made Our Minds Free

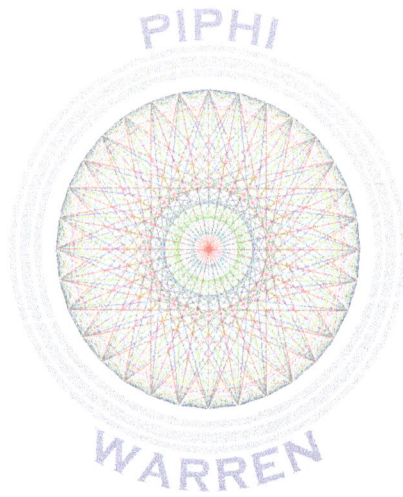

PIPHI

WARREN

TABLE OF CONTENTS

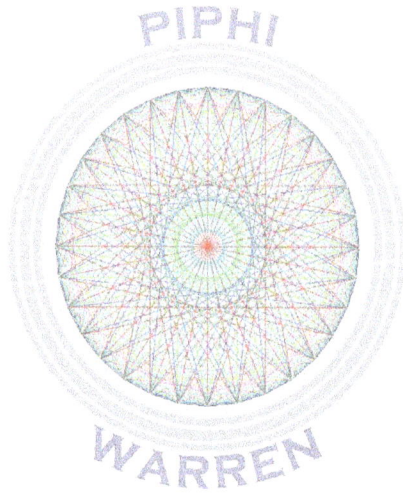

PIPHI WARREN

PREFACE

When I finished "Boscawen-Ûn" sometime in the first months of 2022 I promised myself that I would publish that and my other manuscripts, but that I would write nothing else. After all I was nearly seventy and in perceptible danger of discussing things that I knew nothing about. Or worse, discussing things without knowing that I misunderstood them.

And yet as I have explained elsewhere I am often compelled to research and to write in a way that is often sudden and possibly psychotic. Sadly, unlike my other works, the present book is largely derivative, but to take a more sanguine view that fact may well make it more useful to the university tutors to whom it is addressed than were my former, more personal, enquiries. I hope I have provided adequate references and I apologise to any who have contributed material, but whom I have overlooked.

I hasten to say that virtually any intelligent person can read this book with understanding and interest. There is little calculus and the mathematics never rises much above what used to be O-Level standard, perhaps that to be expected of a studious but not especially gifted seventeen-year-old.

As always there is nothing advanced about my writings but the *issues adumbrated* demand certain levels of intellectual maturity to be considered or even recognised as problems. Accordingly guided learning is sometimes appropriate.

If we are teachers, even of adults, we have to be ever assiduous to avoid projecting our own prejudices or hobby-horses, for our ideas are as mortal as their advocates, and there can be no advance unless the student is encouraged to think for him or herself.

I have had a nodding acquaintance with Pi and Phi for some time, for how long it is difficult to say, but like any aged benefactors their entry to my life was inauspicious, but mysterious and for long unregarded.

My Late Father raised the topic of Pi sometime when I was about nine. Around then my Father was in the protracted process of leaving the Royal Navy. Father was never willing to discuss the details candidly, but it now seems to me that the officers, solicitous to prepare him for civvy life, encouraged him to study the foundations of civil engineering (pun intended, of course), and the elementary mathematics such entailed. At any event, my Father began to study at Camborne Technical College, and when he came home one evening he smilingly raised with me the subject of Pi, apropos what I have no idea and he did not dwell on it at length.

Phi came to me much later. Late one evening I and other foreign postgraduate students were in a waitress-service bar at Bearsden, the only licensed premises in that opulent suburb. A party of men on an adjacent table broke into our conversation to explain to me the rudiments of Golden Ratio theory, why I have no idea, but everyone was very drunk, and I assumed and have continued to believe that they were attempting to make me look stupid. At any event I admitted to knowing nothing about the topic and their leader continued his patronising exposition. The men explained that they worked for Barr and Stroud and this leader of the pack said he doubted I had ever heard of them. I stated that they were rangefinder makers, which much surprised him.

When it comes to an actual "closed form" relationship between Pi and Phi I suppose that the best essay remains Oberg and Johnsons' 2000AD paper "The Pi-Phi Product" which explores the problem in terms of convergent series founded upon geometrical analyses, including such of The Great Pyramid, that remains what its builders intended it to be: A doughty trap for the unwary.

[6 July 2023 note: In the last twenty years Bailey-Borwein-Plouffe (BBP) and Adegoke-Layeni formulations have enabled me to develop functions of the product $\pi\Phi$, not necessarily better than the Oberg-Johnson System.]

If you read any of my previous publications, whether books or offerings on my website "Explorations in the Impossibility of Knowledge" you will recognise my prejudices in the current tome. In particular, I see no distinction between the arts and sciences because if you will forgive the tautology, all human experience is at bottom empirical. Once, at a job interview an Oxford don asked me to define the difference between training and education. I knew the answer he wanted for sure, but I was daft enough to tell him there was no difference. Although I regard myself a Christian I have rightly or wrongly come to despise ideology and dogma, whether sacred or secular, not because of the manifest worldly trouble it causes, but because of the idle and pusillanimous thinking that it betokens.

They say that Science is the new Religion. Both Science and Religion, splendid in themselves, are polluted with numerous superstitions, some fundamental to practice received. To preserve or recover the lovely essences of these glories we must always be on guard, and root out within our minds these blemishes of the surface. Occam's Razor, formulated in a credulous age, is arguably such a superstition, useful with discretion, but impedent of exploration.

Since the War, I mean since 1945AD, some fantastic mathematics books have been written and I have leaned upon them heavily, and also upon some remarkable and indescribably useful software tools such as EXCEL®, WORD®, MathCad®, MicroSoft PhotoDraw®, GIMP®, and many other excellent packages too numerous to recollect, as well as symbolic resources such as Wolfram Alpha® and the ever-reliable Casio Computer Company products.

Wikipedia and its sister sites have also clarified my thoughts on many occasions, in so far as such an improbable thing is possible.

It would be redundant or misleading to list the book titles here, but all errors, and abuses of the resources I have used, are of course mine only.

James R Warren
Midland Tutorial Productions
Bloxwich

16 January 2023

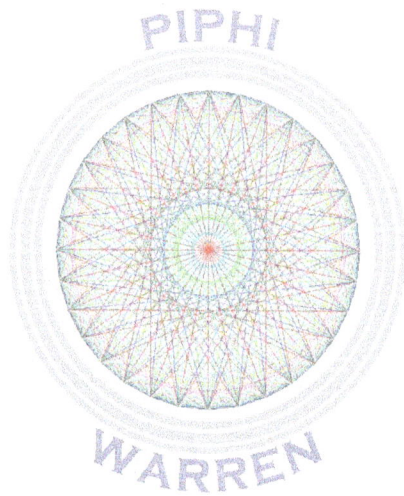

INTRODUCTION

Pi and Phi never married. I hasten to say there is no implication of impropriety in my observation, or in their liaison. Their affinity is quite as honest and as loyal as any relationship can be. For sure, like most of us, Pi fell amongst several scandals in his heady youth, and begat a long-forgotten progeny. But blame him not, for the error was not his, and posterity justifies its own antecedence. Indeed, that progeny, call it what you will, shall forever enjoy his own lusty majority, happy and interesting, but forever bearing the Mark of his illegitimacy.

Pi and Phi exist in a world of Pure Thought, where they quietly lie like the holiest things, not bruting themselves abroad, but awaiting the discovery of a grateful college. Pi and Phi exist in a world of Pure Thought beyond all conjecture, all false reasoning and all other corruptions: Beyond all mortal firmaments that explode into life, dilate and decay to death. Pi and Phi rest content in their own recondite conjugations, more than familiar, but never conterminous. For the destiny of Pi and Phi, like all the finest votaries, is to Describe such evolutions, but not to Decide.

And if the partnership of Pi and Phi is an act of love, then it is an inhuman love, for the love of men is the love of art, and beauty, and the Engineered solution. Pi and Phi know nothing of such things, for as we observed, discretion is not their province, nor their weakness.

It has surprised many that such a private congress has borne issue. Almost secretly, Pi and Phi have a beloved only child, unmasked in Edwardian times. She has no name, though some call her Alpha for want of something more fitting. As with all such issue, vulgar circumlocutions proliferate, to the confusion of honest men.

Alpha is a daughter as fine and as constant as the most fastidious father may wish but hardly structural: Partaking rather of the insubstantial insistences of ghosts, numbers and the holy and unholy Agencies.

As you know there is a certain kind of person who dissembles identity in an effort to be kind or proud, especially to confute the aspersions of the vulgar.

Such afflicted Alpha.

Gossips say that this daughter has nothing to do with Pi and Phi, or even that her parents' espousal is illusory: But we have no need to take cognisance of rude prattling, as we are men and women of refinement, taking the appointments as we find them.

Do not run away with the idea that Pi and Phi are airy-fairy or aethereal abstractions, irrelevant to the commerce of the pavement.

In fact they are like a mundane, perhaps slightly suburban, and if you say so conservative English couple. Though getting on in years, Pi maintains his hobbies of electrodynamics and machine fabrication whilst Phi cleaves to sprightly pastimes involving flowers and the fine arts.

Pi and Phi, long past the age at which rational numbers retire, have workaday jobs to do like anyone else.

Every day of the week, barely pausing for Christmas or Diwali, they take to the planes, and the trains, and other modes of public transport, perfectly together and primly naked in their infinite integuments. For Pi and Phi have no need of clothes, pinching not in the Boreal cold, nor having any need of concealment or deception. At stations, they wait or loiter with proprietorial satisfaction to regard classical mechanics and classical architecture. But Pi and Phi attract not the notice of ticket inspectors, the civil authorities or of the paramedical misericordia, for the commonality fail to perceive them, and the clerisy affect not to notice. Only you and I, and a few old men, take interest, whether to censure or furtively to commune.

And when, in Camden or Kensington, the Tube delivers them to their old and well-appointed workshops they resume their labors of forty hundred years offering a medley and mélange of fact free of prejudice to the delight and instruction of the ever youthful generations.

Yes, Life is full of Paradox, and so is Number. The Life of Numbers is reflected in the Numbers of Life, and Growth the mere prelude to Restoration.

How apt the diffident airs and shy demands of unseen mistresses: Powers without Reason, justified in their own subsistence. Undeniably apt and fitting because they must. Not for the arbitration of men or the satisfaction of beastly pragma, but static and complete as the stars themselves, suggestive in their silence, chaste and pregnant.

CHAPTER ONE
PREDICATES

PART ONE
SOME USEFUL EQUIVALENCES

In terms of the circle of unit radius as developed in Figure 2.1 we may obviously state that the Radius, R, is unity and that AO = R in that diagram.

The Perimeter of that circle is given by 2πR where π is the Ludolphine Constant (sometimes called the Archimedes' Constant), an irrational and a transcendental number, whose value is adjacent to 3.14159265358979

π (pronounced "pie") is the ratio of the Circumference of a Circle to its Diameter (2R) and may succinctly be defined by the Weierstrass Integral as:

$$\pi = \int_{-1}^{+1} \frac{1}{\sqrt{1-x^2}} \approx 2 \int_{-1}^{+1} \frac{1}{1+x^2} \approx 3.14159265358979$$
Equation 1.1

Also, we shall concern ourselves with the Major Ratio of Phidias, sometimes called the Golden Section or other things (such as "PHI" pronounced "fie" or sometimes "fee"), and which is defined by:-

$$\Phi = \frac{1+\sqrt{5}}{2} \approx 1.61803398874989$$

Equation 1.2

The Ratio of Phidias is also irrational.

A closely-associated parameter is the Minor Ratio of Phidias, ϕ, which is Φ-1 defined by:-

$$\phi = \Phi - 1 = \frac{\sqrt{5}-1}{2} \approx 0.61803398874989$$
Equation 1.3

Clearly, the Minor Ration of Phidias must also be irrational.

Often we shall see variously in our derivations the so-called Pythagoras' Constant, Ψ, the square root of two. This is also an irrational number, though technically neither this nor the Ratio of Phidias are transcendental because they are potential roots of algebraic (polynomial) equations.

So:-

$$\sqrt{2} \approx 1.4142135623731$$
Equation 1.4

None of these numbers can be specified *exactly*, but all can be approximated with arbitrary precision, even in terms of each other as we shall show.

Long before we commenced our current investigations it was well-known that[1.1]:-

$$\frac{\pi}{5} = \cos^{-1}\left(\frac{\Phi}{2}\right)$$
Equation 1.5

whilst it is also common knowledge that[1.2]:-

$$\sin^{-1} z = \int_0^z \frac{1}{\sqrt{1 - z^2}} \, . \, dz$$
Equation 1.6

also:-

$$\sin^{-1} z = 2 \tan^{-1}\left(\frac{z}{1 + \sqrt{1 - z^2}}\right)$$
Equation 1.7

and that:-

$$\pi = 2(\sin^{-1} z + \cos^{-1} z)$$
Equation 1.8

so that it is clear to see that:-

$$\cos^{-1} z = \frac{\pi}{2} - \int_0^z \frac{1}{\sqrt{1-z^2}} \, . \, dz$$
Equation 1.9

also:-

$$\cos^{-1} z = 2 \tan^{-1} \left(\frac{\sqrt{1-z^2}}{1+z} \right)$$
Equation 1.10

None of the integrals have a non-recursive analytic solution.

Robert Everest's Equations

$$\Phi = 1 - 2 \cos \left(\frac{3\pi}{5} \right)$$
Equation 1.11

$$\phi = 2 \cos \left(\frac{2\pi}{5} \right)$$
Equation 1.12

Some Simple Functional Relations

Allow that $h = \pi/2$ and $z = \Phi/2$. Then:-

$$\cos^{-1}(z) = \tan^{-1}\left(\frac{\sqrt{1-z^2}}{z}\right) = \tan^{-1}\left(\frac{\sqrt{4-\Phi^2}}{\Phi}\right) = \text{atan2}\left(z, \sqrt{1-z^2}\right)$$

Equation 1.13

whilst:-

$$\cos^{-1}(z) = \tan^{-1}\left(\frac{\sqrt{4-4z^2}}{2z}\right)$$

$$= \tan^{-1}\left(\frac{\sqrt{4(1-z^2)}}{2z}\right) = \tan^{-1}\left(\frac{2\sqrt{(1-z^2)}}{2z}\right)$$

$$= \tan^{-1}\left(\frac{\sqrt{1-z^2}}{z}\right)$$

Equation 1.14

If:-

$$x = \frac{\ln[5\cos^{-1}(z)]}{\ln(z)}$$

Equation 1.15

then:-

$$\pi = \Phi^x$$

Equation 1.16

and:-

$$\pi = 2\left[\int_0^z \frac{1}{\sqrt{1-z^2}} \cdot dz + \cos^{-1}(z)\right] = 2\int_0^z \frac{1}{\sqrt{1-z^2}} \cdot dz + 2\cos^{-1}(z)$$

Equation 1.17

The Right Angled Triangle Not Helpful

Allow that locally the Adjacent of a right-angled triangle is A; the Opposite B = z; and accordingly that the Hypotenuse H = (A²+B²)^(1/2). Then:-

$$\sin^{-1}\left(\frac{B}{H}\right) = \cos^{-1}\left(\frac{A}{H}\right) = \tan^{-1}\left(\frac{B}{A}\right) \neq \cos^{-1}(z)$$

Equation 1.18

PART TWO
SOME USEFUL FUNCTIONS

The Percentage Specific Defect, PSD(x,y)

First of all, we shall according to our usual custom define the (Percentage) Specific Defect, PSD(x,y), as:-

$$PSD(x, y) = 100 \left(\frac{x - y}{x} \right)$$

Equation 1.19

As remarked elsewhere, PSD is closely-related to RMS error and to standard deviation, but it is the simplest possible relative comparator of two outcome values and preserves the sign of the deviation.

If x is the theoretically fiducial value of the parameter under test and y the value found by experiment or approximation, then the PSD gives an immediate sense of the quality of the result.

Fibonacci Successor Functions

In terms of ϕ and Φ we may define successive Fibonacci Sums in terms of:-

$$BSE\phi(n) = \phi^{n-1} + \phi^{n-2}$$
Equation 1.20

$$BSE\Phi(n) = \Phi^{n-1} + \Phi^{n-2}$$
Equation 1.21

The Binomial (Combinatorial) Coefficient

$$NCR(n, k) \equiv BC(n, k) = \frac{n!}{k! \, (n - k)!} = \binom{n}{k}$$
Equation 1.22

The Continuity Divisor Function

$$CONT(n, m) = \frac{\prod_{k=m+1}^{n} k}{\prod_{j=1}^{m} j}$$

Equation 1.23

The Continuity Divisor Function rapidly becomes very large, but is always rational.

The Argument Divisor Function

$$ZSER(z, m) = \sum_{n=0}^{m} \frac{z^{(2n+1)}}{2n + 1}$$

Equation 1.24

The Error Function

The Error Function is defined as:-

$$ERF(z) = \frac{2}{\sqrt{\pi}} \int_{0}^{z} e^{-t^2} . dt$$

Equation 1.25

Selected Trigonometric Functions

Taylor Series Approximations

$$\arcsin(z, m) = \sum_{n=0}^{m} \frac{(2n)!}{[(2^n)n!]^2} \cdot \frac{z^{(2n+1)}}{2n + 1}$$

Equation 1.26

$$\arccos(z, m) = \frac{\pi}{2} - \sum_{n=0}^{m} \frac{(2n)!}{[(2^n)n!]^2} \cdot \frac{z^{(2n+1)}}{2n + 1}$$

Equation 1.27

Inverse Cosine given Inverse Tangent

$$AC(z) = \frac{\sqrt{1 - z^2}}{1 + z}$$

Equation 1.28

using which:-

$$\cos^{-1} z = 2 \tan^{-1}[AC(z)]$$

Equation 1.29

CHAPTER TWO
PI

HIS IMPORTUNATE YOUTH
TRYING HIS STRENGTH WITH HEROS

The pursuit of Pi is littered with the lives of geniuses and Archimedes of Syracuse was arguably the greatest of Europe's men of mind, fit to mention in the same breath as Newton, Einstein and Shakespeare.

When I say the pursuit of Pi I mean that we may run but we may never catch him, because Pi is an incommensurable quantity, not *exactly* knowable.

But it is a fine sport. It is a lot better than the games of my school days standing semi-naked on a vast Hertfordshire field in the keening easterly winds of midwinter. And inarguably a lot more useful.

When we try to relate Phi to Pi it obviously helps to have the best estimate of both figures that we can manage to obtain.

Basically, there are two types of mathematical ways of capturing a number: The "closed form" solution, and the iterative solution.

A "closed form" solution is a single static equation in which you plug known quantities into the right-hand side (RHS) and when you do the arithmetic the solution pops-out fully formed on the left-hand side (LHS).

An iterative solution repeats the same patten of calculations again and again until you reach an answer that satisfies.

But in mathematics, as in life, things as seldom that clear cut, as we shall see.

So what is Pi anyway?

Pi (π) is the ratio of the length of a circle's circumference to its diameter. This ratio is the same for any circle, defined as a two-dimensional figure of constant radius.

There are technical definitions available, such as the Weierstrass Integral, and the Baltzer-Landau Criterion.

Pi is given a wide variety of suitable and unsuitable names in the English Language. It is most often called the Archimedes Constant or the Ludolphine Constant. Both Ludolph van Ceulen and Archimedes of Syracuse were geniuses of former times who contributed to the refinement of the numerical definition of Pi. Mathematics is the art of systematic generalisation. An artist can choose his tools at will, and his subjects too. Van Ceulen and Archimedes, separated by centuries and fame, both did this mathematically and systematically, choosing one each of the infinity of methods that Someone has provided to us.

Because you see, no-one can know the exact value of Pi because a circle's diameter is incommensurable with its perimeter.

Every mathematician worth his salt has had a go at adding extra digits to the definition of Pi. By the Bronze Age, advanced civilisations of the time such as Egypt, Greece, India and China knew that Pi was at least roughly 22/7, a rational fraction with a decimal value of about 3.142857142857142857.....

You can see that from the sixth of the digits after the decimal point the batch of six digits keeps repeating *without adding extra information*. So if we want more accuracy we must find a way of adding more information.

Heroes and geniuses over the millennia have added more information. With regard to Pi, names we can drop include Archimedes, Euler, Gauss, Liu Hu, Weierstrass, Viète, Madhava, Machin, Shanks, Fibonacci, Ramanujan; the list is almost as endless as Pi himself.

The digits of Pi do not actually repeat in any statistically-valid pattern: They are random. This implies that Pi is an Irrational Number so that any actual fraction like 22/7 cannot possibly be a faithful representation of Pi. Because Pi is not the root of any possible algebraic polynomial it is also Transcendental. You may think this rather rare if not arcane, but there are many more Irrational Numbers than Rational Numbers, and probably an infinite number of Transcendentals. So you might think we have our work cut out, but things are by no means as bad as they seem.

So Why Know This?

Firstly I will give you the least important reasons: Those bearing upon science and technology. I cannot do any better than copy wholesale the list of applications given by Wikipedia, out and beyond the best educational resource on the Net, where there are some truly brilliant sites of all kinds:-

Outside mathematics

Describing physical phenomena

Although not a [physical constant](#), π appears routinely in equations describing fundamental principles of the universe, often because of π's relationship to the circle and to [spherical coordinate systems](#). A simple formula from the field of [classical mechanics](#) gives the approximate period T of a simple [pendulum](#) of length L, swinging with a small amplitude (g is the [earth's gravitational acceleration](#)):[194]

$$T \approx 2\pi \sqrt{\frac{L}{g}}$$

One of the key formulae of [quantum mechanics](#) is [Heisenberg's uncertainty principle](#), which shows that the uncertainty in the measurement of a particle's position (Δx) and [momentum](#) (Δp) cannot both be arbitrarily small at the same time (where h is [Planck's constant](#)):[195]

$$\Delta x \Delta p \geq \frac{h}{4\pi}$$

The fact that π is approximately equal to 3 plays a role in the relatively long lifetime of [orthopositronium](#). The inverse lifetime to lowest order in the [fine-structure constant](#) α is[196]

$$\frac{1}{\tau} = 2\frac{\pi^2 - 9}{9\pi} m\alpha^6$$

where m is the mass of the electron.

π is present in some structural engineering formulae, such as the [buckling](#) formula derived by Euler, which gives the maximum axial load F that a long, slender column of length

L, [modulus of elasticity](#) E, and [area moment of inertia](#) I can carry without buckling:[197]

$$F = \frac{\pi^2 E I}{L^2}$$

The field of [fluid dynamics](#) contains π in [Stokes' law](#), which approximates the [frictional force](#) F exerted on small, [spherical](#) objects of radius R, moving with velocity v in a [fluid](#) with [dynamic viscosity](#) η:[198]

$$F = 6\pi \eta R v$$

In electromagnetics, the [vacuum permeability](#) constant μ_0 appears in [Maxwell's equations](#), which describe the properties of [electric](#) and [magnetic](#) fields and [electromagnetic radiation](#). Before 20 May 2019, it was defined as exactly

$$\mu_0 = 4\pi \times 10^{-7} \ H/m$$

A relation for the [speed of light](#) in vacuum, c can be derived from Maxwell's equations in the medium of [classical vacuum](#) using a relationship between μ_0 and the [electric constant (vacuum permittivity)](#), ε_0 in SI units:

$$c = \frac{1}{\sqrt{\mu_0 \varepsilon_0}}$$

Under ideal conditions (uniform gentle slope on a homogeneously erodible substrate), the [sinuosity](#) of a [meandering](#) river approaches π. The sinuosity is the ratio between the actual length and the straight-line distance from source to mouth. Faster currents along the outside edges of a river's bends cause more erosion than along the inside edges, thus pushing the bends even farther out, and increasing the overall loopiness of the river. However, that loopiness eventually causes the river to double back on itself in places *and "short-circuit", creating an <u>ox-bow lake</u> in the process. The balance between these two opposing factors leads to an average ratio of π between the actual length and the direct distance between source and mouth.[199][200]*

In the Seventies of the last century I wrote my PhD on the mathematical analysis of river meandering so I really should have known about the convergence of sinuosity to Pi: But I did not, and at the age of seventy I have learned it for the first time.

Now for the thing really hard to explain.

I think I will start with a little story.

"Once upon a time, a hundred years ago, a young Birkenhead man, a priest's son, strode into The Explorer's Club for a fund-raising press conference..."

No, no, no: This just won't do. Wade Davis is a much better storyteller than I. Let him take up the tale[2.1]:-

"Mallory returned to New York in a restless mood, anxious to get home. A failed attempt to climb a mountain evidently could not capture an American imagination fired by more splendid news, such as the discovery in Egypt of the treasure of King Tutankhamen, a story that broke in the New York papers on February 17, or the Tibetan adventures of William Montgomery McGovern, a Buddhist scholar who sneaked into Lhasa disguised as a monk, his skin and hair blackened with dye, only to be imprisoned and forced to make a harrowing escape. The only thing about Everest that seemed of interest to the American press was the fact that Mallory had taken a swig of brandy at 27,000 feet. Booze was a local hook. Prohibition was just beginning to convulse the country.

*Oddly enough, the most memorable note to turn up in the American papers was a casual remark Mallory made at the end of one of his lectures. Asked why he wanted to climb Everest no doubt for the umpteenth time, Mallory reportedly replied, "**Because it's there**". This simple retort hit a nerve, and took on an almost metaphysical resonance, as if Mallory had somehow in his wisdom distilled the perfect notion of emptiness and pure purpose. It was first quoted in the Sunday New York*

Times on March 18, in the opening paragraph of a half-page feature, "Climbing Mount Everest Is Work for Supermen". In time, it would be inscribed on memorials, quoted in sermons, cited by princes and presidents. But those who knew Mallory best, including two of his biographers, his close friend David Pye and his son-in-law David Robertson, interpreted the comment rather more casually. To them it was simply a flippant response by an exhausted and frustrated man who famously did not suffer fools. Or as Arnold Lunn remarked, Mallory, no stranger to New York speak-easies just said it to get rid of 'a bore who stood between him and a much needed drink'.

Whatever its genesis, the phrase caught on because it did in fact capture something essential. 'Everest is the highest mountain in the world', Mallory later wrote, 'and no man has reached its summit. Its existence is a challenge. The answer is instinctive, a part, I suppose of man's desire to conquer the universe'. Elsewhere he added, 'I suppose we go to Mount Everest, granted the opportunity, because—in a word—we can't help it. Or to state the matter rather differently, because we are mountaineers."

Now I hate all sports. As I implied earlier, like many other Englishmen, I was forever immunised at school. I am especially disdainful of alpinism, for me the activity of climbing mountains, (unless there is a clear scientific or military end in view), is the very epitome of expensive and dangerous pointlessness, worse than racing, better than hunting.

But I neither disdain nor despise climbers. In my crabby, vermicular way I respect and even admire what they try to do. For courage and ambition are admirable, inspiring, worshipful, especially when they harm neither man nor beast.

Courage and ambition is what is needed when you fight to extend Pi. It is not a matter of intellect, or learning, or even patience. It is a matter of being a Hero, a man or a woman who loves, and loves something ardently enough to sacrifice the Self for it. It is beyond cash, safety or fame. So if Archimedes, Viète, Ramanujan and the rest were Heroes, then who are the cowards? The cowards are the people who excuse their indolence or lack of resolve with a sneaking abdication to the security of self-exculpation, turning their backs upon peril, glory and enterprise. A coward experiments upon small creatures because he cannot bear to pain himself, not even by the hard but fecund work of analysis and simulation. The coward vanishes into the mass, blames the failures of his fellows, pretends to deprecate success, and regards with jealous snobbery the ever-unattainable good objectives of heroes. The coward is politics perfected.

When I was a boy I considered the phrase "Because it's there" the unimprovable confession of banal vacuity. It also seemed it had more than a trace of nihilism. Now I am a man I know that men must strive to achieve, that resignation of achievement is Death, and that the confession of love is Life. Yes, even if you know that what you seek is unhaveable such as an exact solution for Pi or Phi.

Real men and real women struggle. Real people struggle to understand. Real men struggle to attain the summit, literally or figuratively. Real people struggle to approach The Cross.

But, but, but...there is always a "but" in life, isn't there?

Objectives beguile, especially when they are, or are thought to be, unapproachable. The summit of Everest is for ever swept by a freezing hurricane of rarefied gas that eagles would dare if only they could grip the air. Like the last pit of Dante's Inferno its ice is perpetual, the slightest slip offering a twenty-thousand foot drop on each side. Its allure is irresistible. The irrepudiable romance of the inaccessible.

Many hundreds of men and women have died unhallowed upon the slopes of Everest. We only know of George Mallory and his climbing partner Andrew "Sandy" Irvine (by coincidence another Birkenhead man) because they *possibly* succeeded in accomplishing what they set out to do, and *might* have done so thirty years before Hillary and Tensing were *known* to have summited the fatal hill.

Even if you are a very young person I am sure that you understand that Everest, the Moon, Mars, Fermat's Last Theorem, the Squared Circle, socialism, The Conservative Party and indeed the girl or boy at the back of the class who you always fancied are all "almost metaphysically resonant" things. And like all metaphysical adversaries without real puissance they are psychospiritual perils lying in wait to drive the Hero to madness and even early death.

For the weak or merely unfortunate mathematician or other explorer is somewhat in the predicament of the man who courts ghosts or even tries to exorcise them (I speak literally or figuratively at your pleasure, and I hope, at your service).

"Almost metaphysically resonant" things have a glamor all their own and men and women invest them with an idolatrous allure that the objects themselves neither seek nor incite so that with every step taken to the goal their projected power becomes twice as great, until eventually the seeker destroys his own fortune, health and mortal life.

Of course, this has been the subject of great literature of all nations, and every age. But in this book I need to stick to mathematics.

So how did Archimedes attack Pi?[2.2]

Archimedes was a Syracusan. At this point in time we are not sure whether he saw himself as a Syracusan or a Greek. Much more recently people who lived in America were not sure whether they were American or British. The great poet, TS Eliot, was possibly among the last such people.

One thing Archimedes was sure about: He was not a Roman.

He was a loyal Syracusan, allegedly related to the king. When the Roman fleet besieged Syracuse in 213BC Archimedes set to designing weapons for use against the Roman ships, armaments like heat rays and grappling cranes, centuries ahead of their time.

It is Archimedes we must thank for our knowledge of the circle, the sphere and the arithmetic spiral; for our knowledge of hydrostatic buoyancy and many other scientific principles.

Like many other people, including laymen, I have long known that if you overload a bridge it will surely collapse, a fact attested by scores of sad disasters. Now I am a trained hydraulician, but I learned only the other day that if you pass a vessel of any tonnage along a canal and over an aqueduct you will not add to the stress upon the structure, because the water and weight displaced distributes to the waterway *along the two approaches* to the bridge. So if an aqueduct is going to collapse, it will do so under its own unburdened self-weight.

The Romans wanted Archimedes, perhaps loved him. Plutarch relates that after Marcellus had occupied the city he ordered soldiers to arrest the great mathematician and bring him for interview, but on no account to harm him. The detail found him hard at work drawing diagrams in his study. They ordered him to come with them. Archimedes replied "Do not disturb my circles". He was slain on the spot. Marcellus was livid.

Archimedes realised, quite when is lost to history, that if you progressively increased the number of (finite) sides of a regular polygon it would as a figure approximate ever more closely a circle.

Now my approach, with the benefit of modern trigonometry and of course my computer, would be to calculate and total the individual, but identical, sides of the polygon and compare the length obtained with its precursors.

Archimedes approach was different: Archimedes considered the length of the perimeters of successive inscribed and circumscribed circles for successive polygons. Archimedes approach was much more intelligent (and economic) than my approach, because it converges to an accurate estimate much more rapidly. The geometric situation is illustrated by me below:-

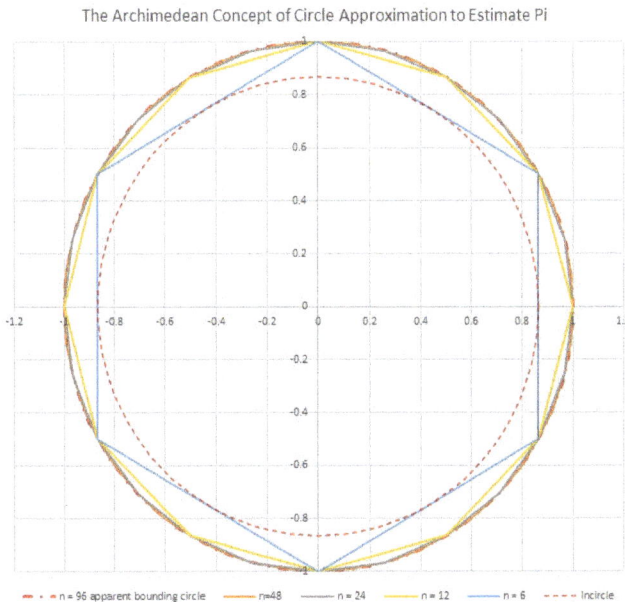

The Archimedean Concept of Circle Approximation to Estimate Pi

Figure 2.1
The Archimedean Approach to Computing a Value for Pi

Notice two important things:-

A   Archimedes *doubled* the number of polygon sides at each iteration, starting with a Hexagon (n = 6)

B   Archimedes did not ask himself about
the Polygon Sides.
Archimedes asked the question
"What are the relative areas of circles
inscribed and
circumscribed about the relevant polygon
at each iteration"

For the n = 6 hexagon my diagram shows the incircle as red dashes and the circumcircle as red dash-dots.

What this boils down to at each stage is finding the Area of the 6n'th Annulus:-

$$A_{6n} = \pi\left(R_{6n}^2 - r_{6n}^2\right)$$
Equation 2.1

where A_{6n} is the Annulus Area, R_{6n} is the radius of the Circumcircle (you can maintain this constant for all the polygons at a value of unity); and R_{6n} is the Iteration Incircle Radius. At all stages the following applies:-

$$A_{incircle} < A_{6n} < A_{circumcircle}$$
Inequality 2.1

and:-

$$P_{incircle} < P_{6n} < P_{circumcircle}$$
Inequality 2.2

Pi is of course not known at this time, but knowing by definition that:-

$$P = \pi.2\rho$$
Equation 2.2

where P is the Circle Perimeter and ρ is the Circle's Radius it becomes possible to infer a value for Pi.

Starting at n = 6 and working up to n = 96 Archimedes used the following patterns of iterated square roots which I give for the case of the 24-gon:-

$$LB < \pi_{estimated} < UB$$
Inequality 2.3

$$LB_{4\times6} = 24 \times \left(\frac{\sqrt{2 - \sqrt{2 + \sqrt{3}}}}{2} \right)$$
Equation 2.4

$$UB_{4\times6} = 24 \times \left(\frac{\sqrt{2 - \sqrt{2 + \sqrt{3}}}}{\sqrt{2 + \sqrt{2 + \sqrt{3}}}} \right)$$
Equation 2.5

By the 96th iteration Archimedes knew that the "adequate" value of Pi stood somewhere between the fractions 22/7 and 223/71. In this case LB and UB are estimates of P/(2ρ) which of course is an estimate of Pi itself.

Trigonometry had not been invented when Archimedes worked but in modern trigonometrical terms Archimedes' process is epitomised by the iterated equations:-

$$LB_n = n.\sin\left(\frac{\pi}{n}\right)$$
Equation 2.6a

$$UB_n = n.\tan\left(\frac{\pi}{n}\right)$$
Equation 2.6b

Table 2.1 presents a list of the values of Pi estimates according to the Equations 2.6a and 2.6b system:-

MODERNISED ARCHIMEDES METHOD

Number of Sides	LB(n)	UB(n)	Geometric Mean of LB and UB	PSD $(\pi, \pi_{implied})$
3	2.59807621	5.19615242	3.67423461	-16.95452019
6	3.00000000	3.46410162	3.22370980	-2.61386981
12	3.10582854	3.21539031	3.16013465	-0.59021001
24	3.13262861	3.15965994	3.14611525	-0.14395860
48	3.13935020	3.14608622	3.14271640	-0.03577010
96	3.14103195	3.14271460	3.14187316	-0.00892888
192	3.14145247	3.14187305	3.14166275	-0.00223137
384	3.14155761	3.14166275	3.14161018	-0.00055779
768	3.14158389	3.14161018	3.14159703	-0.00013944
1536	3.14159046	3.14159703	3.14159375	-0.00003486
3072	3.14159211	3.14159375	3.14159293	-0.00000872
6144	3.14159252	3.14159293	3.14159272	-0.00000218

Table 2.1
Archimedes Numerical Values for
Iterates of The Trigonometrical Representation

Viète's Formula[2.3]

Like another great French mathematician, Pierre de Fermat, who worked half a century later, François Viète was a lawyer. Viète was also a wealthy landowner of The Vendée who had queens for clients and indeed acted for Mary Queen of Scots whilst she was in France. It is not known whether Viète pleaded for her in England. He worked confidentially in days of caution.

As you may have adjudged, Viète was very politically active whether he relished it or not, and got into a lot of trouble for his alleged Huguenot (Protestant) sympathies. However, like Archimedes he was too useful to be ordered killed by either side and found especial favor for breaking the Spanish ciphers to the great advantage of King Henri, of France and, indirectly, the French Protestant cause.

Although Viète lived in an age of great algebraists he is best remembered for introducing the custom of using letters to denote variables in equations, a simple and apparently obvious innovation which greatly assisted future thought in mathematics and science.

Very similar in spirit to Archimedes Method is Viète's Formula for Pi which may be represented in modern terms as:-

$$\pi = 2 \times \frac{2}{\sqrt{2}} \times \frac{2}{\sqrt{2 + \sqrt{2}}} \times \frac{2}{\sqrt{2 + \sqrt{2 + \sqrt{2}}}} \times \frac{2}{\sqrt{2 + \sqrt{2 + \sqrt{2 + \sqrt{2}}}}} \times \ldots$$

Equation 2.7

or concisely:-

$$\pi = \frac{2}{\prod_{n=1}^{\infty} \cos\left(\frac{\pi}{2^{(n+1)}}\right)}$$

Equation 2.8

The Father of Modern Algebra, this Frenchman François Viète published his method in 1593AD.

The latter formula, given by Wikipedia, is clearly circular in the sense that it depends upon prior knowledge of Pi.

A stepwise algorithm given on the Keisan Casio[2.4] website follows the spirit of the Viète method without anticipating knowledge of Pi:-.

$s_0 = 0$	**Equation 2.9a**
$y_0 = 1$	**Equation 2.9b**

$s_{k+1} = \sqrt{\frac{1 + s_k}{2}}$	**Equation 2.10a**
$y_{k+1} = y_k \cdot s_{k+1}$	**Equation 2.10b**

$$PI_k = \frac{2}{y_k} \qquad \textbf{Equation 2.11}$$

Please note that $y_{(k+1)}$ is updated using $y_{(k)}*s_{(k+1)}$ and not $y_{(k)}*s_{(k)}$ as quoted on the Keisan website.

As aforestated, we confirm that The Keisan Tabular Method of the Viète Solution does not require prior knowledge of Pi.

When we fix m the Number of Iterations at 16 and stop at $PI_m = 2/y_m$, we have:-

$$\pi_{implied} = PI_m = \frac{2}{y_{16}} = 3.14159265328899$$

$$\textbf{Equation 2.12}$$

For which the relevant Percentage Specific Defect is 0.00000000957477

KEISAN-CASIO VIÈTE ALGORITHM

Number of Sides	s_k	y_k	$2/y_k$	PSD $(\pi, \pi_{implied})$
3	0.92387953	0.65328148	3.06146746	2.55046416
6	0.99879546	0.63687551	3.14033116	0.04015469
12	0.99999971	0.63661983	3.14159235	0.00000980
24	1.00000000	0.63661977	3.14159265	0.00000000

Table 2.2
The Superior Convergence of Viète's Method
As instantiated by the Keisan Casio Algorithm

Ramanujan's Formula

Srinivasa Ramanujan FRS was a great self-taught genius of Tamil heritage. He was the second Indian to be elected a Fellow of the Royal Society, Britain's leading academy of science, and the first to be elected a Fellow of Trinity College, Cambridge, Isaac Newton's alma mater.

Professional mathematicians are very good at telling each other that they have "proved" this or that outcome to be impossible. Ramanujan was not a professional. Ramanujan was a devoutly religious Hindu who wrote that "An equation for me has no meaning unless it expresses a thought of God." Ramanujan worked as a clerk.

He was born at Erode in 1887AD. Erode was a provincial town in Mysore State, a semi-autonomous, semi-enclave nearly surrounded by the Madras Presidency province of British India. His father worked in a ladies' dress shop and his mother was a temple singer. The family lived in great poverty. Srinivasa lost siblings to disease at early ages, and he himself recovered from smallpox as a toddler.

Ramanujan's mathematical abilities were the talk of his contemporaries at school and in the locality at large, and he would often help his teachers with timetabling classes and other thorny logistical, but thankless, tasks.

As a teenager Ramanujan would borrow mathematical textbooks from friends and libraries but gained no formal tuition beyond what he picked up at high school. When sixteen years old he independently re-discovered the Bernoulli numbers, and the solution of quartic equations, and attempted to solve the quintic, not knowing that it was "insoluble" by classical means.

At the time, as he grew to manhood, Srinivasa Ramanujan was unable to further his mathematical interests in India, and so he initiated a correspondence with Godfrey A Hardy, Wykehamist and Smith's Prize holder, a bashful and reclusive mathematician of Trinity College.

In 1913 Hardy received a letter bearing Indian postage stamps and opened it to find pages of dazzlingly-original mathematical theorems and identities.

It would not have occurred to Hardy that an Indian may have been socially-unacceptable in a Cambridge college in 1913AD, haunted as it was by white Anglo-Saxon Protestant ex-Clarendon men. Or, indeed, that a frail man, used to a sub-tropical climate, might be damaged by the damp, cold British airs. So Hardy eagerly invited Ramanujan to Trinity College as a kind of protege. Ramanujan gave up his job as a clerk and jumped on the next boat.

At Cambridge Hardy helped Ramanujan publish 3,900 original equations, and Ramanujan's work provided material for others usefully to develop, well into the twenty-first century.

But already a sick man, Ramanujan developed hepatic amoebiasis, an often fatal digestive disease. Ramanujan returned to India to die, where Hardy met him on his deathbed.[2.5]

> *As the two friends met for the last time Hardy said:-*
> *"I drove over here to-day in a taxi with a boring number, 1729."*
> *Ramanujan's face brightened as though he had met an old friend,*
> *"Oh no, that's not a boring number;*
> *it's the smallest number that's representable*
> *as a sum of two cubes*
> *in two different ways!"*

Srinivasa Ramanujan was thirty-two.
At the end of a long and productive life Hardy was asked:-
"What was your greatest contribution to science?"
Hardy replied:-
"The Discovery of Ramanujan".

Ramanujan's Formula for the Approximation of Pi is:-

$$\frac{1}{\pi} = \frac{2\sqrt{2}}{9801} \cdot \sum_{k=0}^{m} \frac{(4k)!\,(1103 + 26390k)}{(k!)^4 . 396^{4k}}$$

Equation 2.13

where k is the Serial Number of the Iteration, m is the Number of Iterations, and π is obviously The Ludolphine Constant.

This equation is one of the great compositions of World literature. It completely mystifies me how a mortal mind could possibly arrive at it.

Consider the position when m = 0, that is to say there are no iterations, the figures are taken as they stand: "Closed-Form".

Armed with pencil and paper, high-school knowledge of the laws of exponents, and a modicum of patience you can easily check that:-

$$\frac{1}{\pi} = \frac{2\sqrt{2}}{9801} \cdot \sum_{k=0}^{0} \frac{(4k)!\,(1103 + 26390k)}{(k!)^4.\,396^{4k}} = 2\sqrt{2}\,\frac{1103}{9801}$$

Equation 2.14

or:-

$$\pi = 3.14159273001331$$
Equation 2.15

For which the PSD(π,$\pi_{implied}$) is -0.000002432635894 or in other words Ramanujan has already given us Pi to seven figures.

Now my copies of MathCad® Express®, and MicroSoft® EXCEL®, and also my 2021 Dell® XPS 64-bit microcomputer all give me fifteen-figure accuracy. So, being a cunning old cove I would like to make me a nice "closed-form" pseudoanalytic equation for Pi to use to, say, a cool fourteen places of decimals.

(Yes, I really do take a pride in being basic and boring:- Except of course that I never want to bore my esteemed readers!).

Here goes:-

$$\frac{1}{\pi} = \frac{2\sqrt{2}}{9801} \cdot \left[\sum_{k=0}^{0} \frac{(4k)!\,(1103 + 26390k)}{(k!)^4.\,396^{4k}} + \sum_{k=1}^{1} \frac{(4k)!\,(1103 + 26390k)}{(k!)^4.\,396^{4k}} \right]$$

Equation 2.16

or:-

$$\frac{1}{\pi} = \frac{2\sqrt{2}}{9801} \cdot \left[\frac{(4 \times 0)!\,(1103 + 26390 \times 0)}{(k!)^0.\,396^{4 \times 0}} \right.$$
$$\left. + \frac{(4 \times 1)!\,(1103 + 26390 \times 1)}{(k!)^1.\,396^{4 \times 1}} \right]$$

$$\frac{1}{\pi} = \frac{2\sqrt{2}}{9801} \cdot \left[\frac{(0)!\,(1103)}{1.1} + \frac{(4)!\,(1103 + 26390)}{1! \times 396^4} \right]$$

$$\frac{1}{\pi} = \frac{2\sqrt{2}}{9801} \cdot \left[(1103) + \frac{24 \times (1103 + 26390)}{1 \times 396^4} \right]$$

$$\frac{1}{\pi} = \frac{2\sqrt{2}}{9801} \cdot \left[1103 + \frac{24 \times (1103 + 26390)}{1 \times 2.4591257856 \times 10^{10})} \right]$$

$$\frac{1}{\pi} = \frac{2\sqrt{2}}{9801} \cdot \left[1103 + \frac{659832}{24591257856} \right]$$

$$\frac{1}{\pi} = \frac{2\sqrt{2}}{9801} \cdot [1103 + 0.000026831974349]$$

Equation 2.17

From which:-

$$\pi = 3.14159265358979$$
Equation 2.18

For which the PSD($\pi,\pi_{implied}$) is -0.000000000000014 or in other words Ramanujan has already given us Pi to fifteen figures in a mere iteration of his approximate formula.

RAMANUJAN METHOD

Number of Iterations	Ramanujan Sum Term	Ramanujan Sum Cumulate	Ramanujan Estimate $1/\pi_{implied}$	Ramanujan $\pi_{implied}$	PSD $(\pi, \pi_{implied})$
0	1103	1103	0.31830988	3.141592730013310	-2.43264E-06
1	2.6832E-05	1103.000027	0.31830989	3.141592653589790	-1.41358E-14
2	2.24539E-13	1103.000027	0.31830989	3.141592653589790	0
3	1.99507E-21	1103.000027	0.31830989	3.141592653589790	0
4	1.83935E-29	1103.000027	0.31830989	3.141592653589790	0
5	1.73588E-37	1103.000027	0.31830989	3.141592653589790	0

Table 2.3
The Ultra-Rapid Convergence of Ramanujan's Formula

The Classical Methods for Pi: Compared for Estimated Pi

Figure 2.1 is a plot to illustrate the relative convergence of the Side Summation Method; the Modernised Archimedes Method: the Kaisan-Casio Viète Algorithm; and the Ramanujan Method. It can be seen that the Side Summation Method and the Archimedes Method are operationally the same, except that the former converges from below, and the Archimedes method from above.

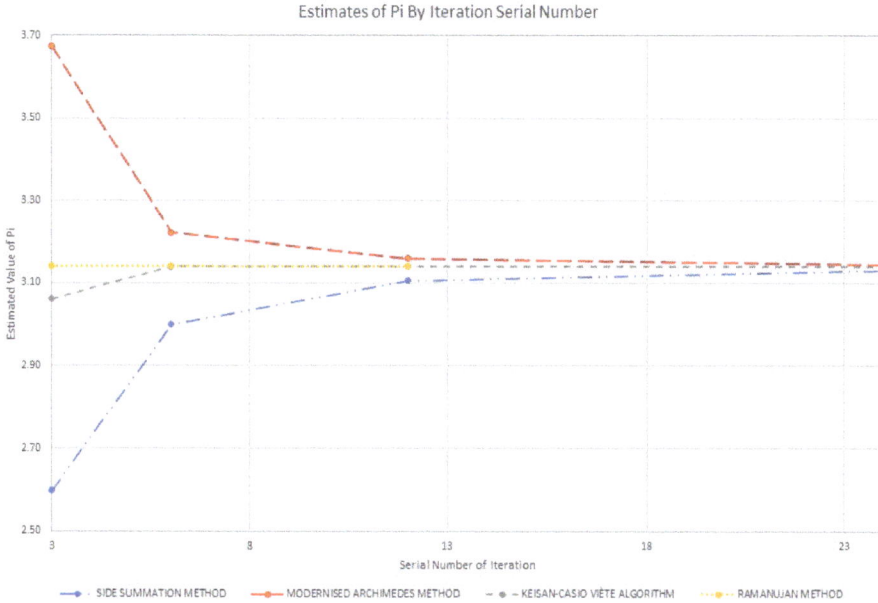

Figure 2.1
The Convergences of Estimates of the Ludolphine Constant π
For Various Classical Iterative Methods

The Classical Methods for Pi: Compared for Specific Defect of Pi Estimates

The Percentage Specific Defect $PSD(\pi,\pi_{implied})$ was computed at each iterate of the Side Summation Method; the Modernised Archimedes Method: the Kaisan-Casio Vi**è**te Algorithm; and the Ramanujan Method.

Figure 2.2 is a plot to illustrate the relative convergence of the Side Summation Method; the Modernised Archimedes Method: the Kaisan-Casio Vi**è**te Algorithm; and the Ramanujan Method in terms of the PSD. It can be seen that the Side Summation Method is about half as good at any iteration as the Archimedes Method but is morphologically very similar. The Viète system converges to fourteen-figure accuracy in four doubled-n iterations, i.e. at n = 3,6,12,24.

Ramanujan's Method has already hit fifteen-figure fidelity before n = 3 is reached.

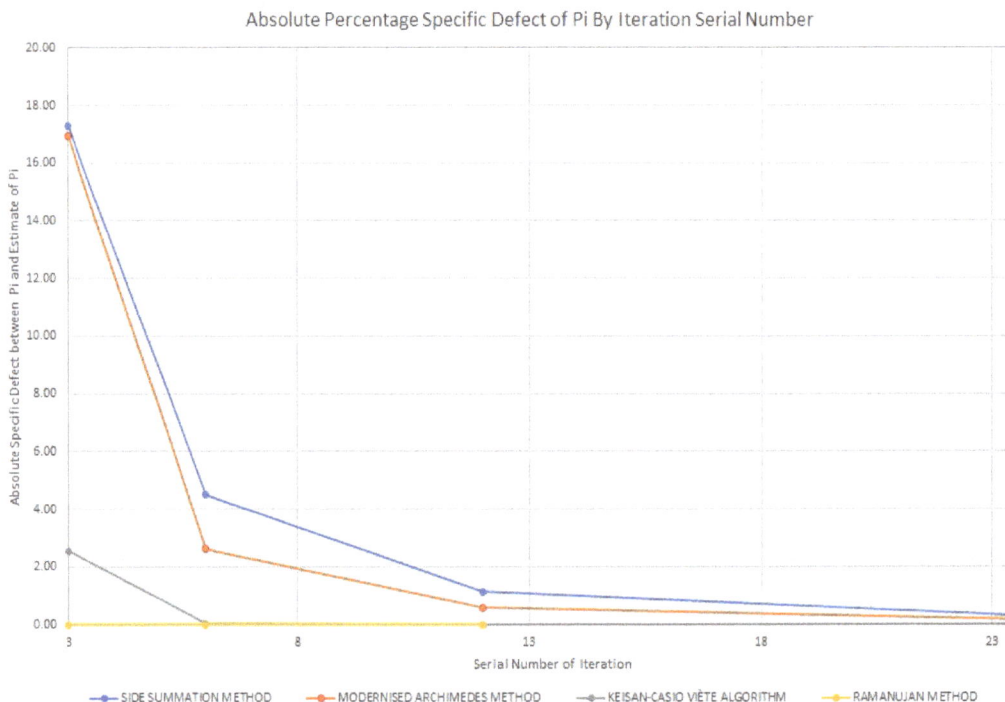

Figure 2.2
The Convergences of PSD($\pi,\pi_{implied}$) Precision Metrics
Of the Ludolphine Constant π
For Various Classical Iterative Methods

I am sure you would forgive a student for thinking there was no more to be said about Pi.

But of course he would be wrong!

The Regular Decagon[2.6]

Figure 2.3 illustrates the regular decagon with ten red sides as continuous lines. The blue dotted line is its circumcircle which I mostly included in order to adjudge co-equality of extension in the EXCEL® diagram.

Two adjacent vertex radii are shown as feint green dashed lines and their included angle is γ, which is of course 36°, and is in radians:-

$$\gamma = \frac{2\pi}{n} = \frac{2\pi}{10} = \frac{\pi}{5}$$

Equation 2.19

n = 10 because the decagon has ten sides, and each side has Length s.

Polygon Plot

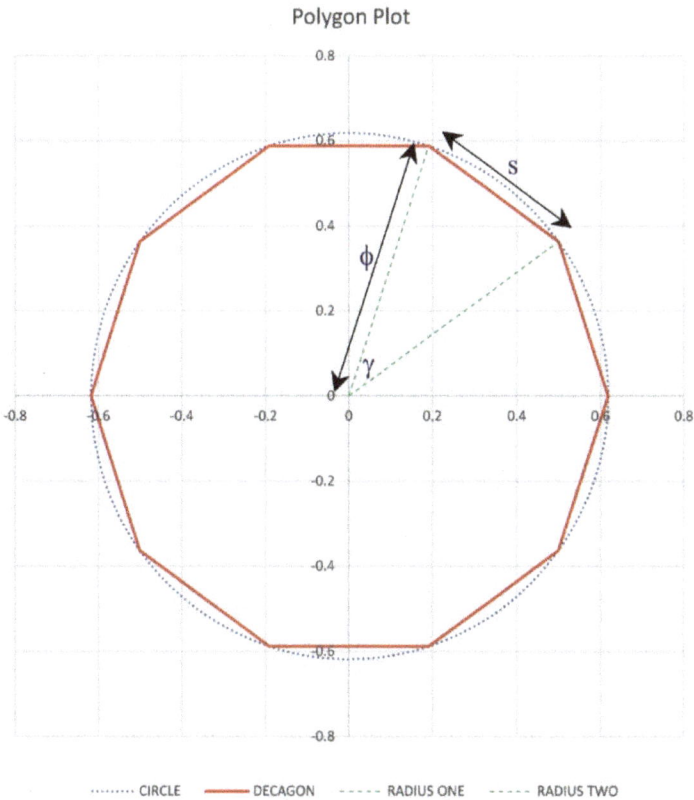

Figure 2.3
The Regular Decagon

Now in our usual manner we shall confirm that:-

$$\Phi = \frac{1 + \sqrt{5}}{2} = \phi + 1$$

Equation 2.20

and note that for the Regular Decagon:-

$$\frac{\Phi}{1} = \frac{\phi}{s} = 1.61803398874989$$

Equation 2.21

and for that Regular Decagon it follows that:-

$$\cos(\gamma) = \frac{1}{2\phi} = 0.809016994374947$$

Equation 2.22

The Vertex Radius, R, is ϕ; so s is 0.381966011250105
By the Cosine Rule:-

$$s = \sqrt{\phi^2 + \phi^2 - 2.\phi.\phi.\cos\left(\frac{1}{2\phi}\right)} = \sqrt{2\phi^2 - \phi}$$

Equation 2.23

and the Area of the Triangle Subtending γ and its opposite Side s, A_T, is given by:-

$$A_T = \frac{1}{2}.\phi.\phi.\sin(\gamma)$$

$$= \frac{\phi^2}{2}.\left[\sqrt{1 - \left(\frac{1}{2\phi}\right)^2}\right]$$

$$= \frac{\phi^2}{2} \cdot \left[\sqrt{1 - \frac{1}{4\phi^2}} \right]$$

Equation 2.24

This is because the Area of this Regular Decagon, A_D, is given by:-

$$A_D = 5\phi^2 \cdot \sin\left(\frac{\pi}{5}\right)$$

Equation 2.25

from which it evidently follows that:-

$$A_T = \frac{1}{2}\phi^2 \cdot \sin\left(\frac{\pi}{5}\right)$$

Equation 2.26

Therefore, by equivalating Equations 2.24 and 2.26 we have:-

$$\frac{\phi^2}{2} \cdot \left[\sqrt{1 - \frac{1}{4\phi^2}} \right] = \frac{1}{2}\phi^2 \cdot \sin\left(\frac{\pi}{5}\right)$$

Equation 2.27

and:-

$$\left[\sqrt{1 - \frac{1}{4\phi^2}} \right] = \sin\left(\frac{\pi}{5}\right)$$

Equation 2.28

Therefore:-

$$\cos\left(\frac{\pi}{5}\right) = \sqrt{\frac{1}{4\phi^2}} = \frac{1}{2\phi}$$

Equation 2.29

as previously shown.

A Complex Treatment using Euler's Formula

Now by Euler's Formula:-

$$e^{i\frac{\pi}{5}} = \cos\left(\frac{\pi}{5}\right) + \sin\left(\frac{\pi}{5}\right).i$$
Equation 2.30

where i is the Square Root of Minus One, $-1^{\frac{1}{2}}$, so Equation 2.30 is complex.

In numerical terms:-

$$e^{i\frac{\pi}{5}} = \cos\left(\frac{\pi}{5}\right) + \sin\left(\frac{\pi}{5}\right).i$$
$$= 0.809016994374947 + 0.587785252292473i$$
Equation 2.31

from which it follows that:-

$$i\frac{\pi}{5} = \ln(0.809016994374947 + 0.587785252292473i)$$
$$= 0.628318530717959i$$
Equation 2.32

Furthermore:-

$$\left(e^{i.\frac{\pi}{5}}\right)^2 = 0.309016994374947 + 0.951056516295154i$$
Equation 2.33

or:-

$$e^{i.\frac{2\pi}{5}} = 0.309016994374947 + 0.951056516295154i$$
Equation 2.34

Note that:-

$$\cos\left(\frac{\pi}{5}\right) - \frac{1}{2} = 0.309016994374947$$
Equation 2.35

and that:-

$$\sin\left(\frac{2\pi}{5}\right) = 0.951056516295154$$
Equation 2.36

by substitution we may declare:-

$$e^{i\cdot\frac{2\pi}{5}} = \left[\cos\left(\frac{\pi}{5}\right) - \frac{1}{2}\right] + \left[\sin\left(\frac{2\pi}{5}\right)\right]i$$
Equation 2.37

Now Sin(2π/5) may be converted to a surd and simplified in these terms:-

$$\sin\left(\frac{2\pi}{5}\right) = \sqrt{\frac{5 + \sqrt{5}}{8}}$$

$$= \frac{1}{2}\cdot\sqrt{\frac{5}{2} + \frac{\sqrt{5}}{2}}$$

$$= \frac{1}{2}\cdot\sqrt{\frac{5 + \sqrt{5}}{2}}$$

$$= \frac{1}{2}\cdot\sqrt{\frac{1 + \sqrt{5}}{2} + \frac{4}{2}}$$

$$= \frac{\sqrt{\Phi + 2}}{2}$$
Equation 2.38

In numerical terms:-

$$\sin\left(\frac{2\pi}{5}\right) = 0.951056516295154$$
Equation 2.39

as above.
Meanwhile:-

$$\cos\left(\frac{\pi}{5}\right) - \frac{1}{2} = \left(\frac{1+\sqrt{5}}{4}\right) - \frac{1}{2}$$

$$= \frac{\sqrt{5}-1}{4}$$

$$= \frac{\phi}{2}$$
Equation 2.40

By substitution in Equation 2.37 we attain:-

$$e^{i\cdot\frac{2\pi}{5}} = \frac{\phi}{2} + \frac{\sqrt{\Phi+2}}{2}i$$
Equation 2.41

or:-

$$e^{i\cdot\frac{2\pi}{5}} = 2\left(\phi + \sqrt{\Phi+2}i\right)$$
Equation 2.42

which by co-ordination on ϕ is:-

$$e^{i\cdot\frac{2\pi}{5}} = 2\left(\phi + \sqrt{\phi+3}i\right)$$
Equation 2.43

A Real Approximation

In numerical terms:-

$$e^{-1 \cdot \left(\frac{\pi}{5}\right)^2} = e^{-\left(\frac{\pi^2}{25}\right)} = 0.673825451231434$$
Equation 2.44

Also by reference to Equation 2.26 for the Area of the Triangle subtending $\pi/5$:-

$$A_T = \frac{1}{2}\phi^2 \cdot \sin\left(\frac{\pi}{5}\right)$$
Equation 2.26

we obtain:-

$$6 \cdot A_T = 6\left[\frac{1}{2}\phi^2 \cdot \sin\left(\frac{\pi}{5}\right)\right] = 3\phi^2 \cdot \sin\left(\frac{\pi}{5}\right) = 0.673541964869378$$
Equation 2.45

which approximately equivalates Equation 2.44, so that:-

$$e^{-1 \cdot \left(\frac{\pi}{5}\right)^2} = e^{-\left(\frac{\pi^2}{25}\right)} \approx 3\phi^2 \cdot \sin\left(\frac{\pi}{5}\right)$$
Equation 2.46

Since in surd form $\mathrm{Sin}(\pi/5)$ is given by:-

$$\sin\left(\frac{\pi}{5}\right) = \sqrt{\frac{5 - \sqrt{5}}{8}} = \sqrt{1 - \frac{1}{4\phi^2}}$$
Equation 2.47

we may write:-

$$e^{-1.\left(\frac{\pi}{5}\right)^2} = e^{-\left(\frac{\pi^2}{25}\right)} \approx 3\phi^2 . \sqrt{1 - \frac{1}{4\phi^2}}$$

Equation 2.48

Trigonometric Ratios

We established above that:-

$$\cos\left(\frac{\pi}{5}\right) = \frac{1}{2\phi}$$

Equation 2.49

From which it follows that because:-

$$\left[\cos\left(\frac{\pi}{5}\right)\right]^2 + \left[\sin\left(\frac{\pi}{5}\right)\right]^2 = 1$$

Equation 2.50

it is manifest that:-

$$\sin\left(\frac{\pi}{5}\right) = \sqrt{1 - \frac{1}{4\phi^2}}$$

Equation 2.51

Accordingly:-

$$\cos\left(\frac{\pi}{5}\right) + \sin\left(\frac{\pi}{5}\right) = \frac{1}{2\phi} + \sqrt{1 - \frac{1}{4\phi^2}} = 1.39680224666742$$

Equation 2.52

To establish the Tangent of $\pi/5$ we may write:-

$$\tan\left(\frac{\pi}{5}\right) = \frac{\sin\left(\frac{\pi}{5}\right)}{\cos\left(\frac{\pi}{5}\right)} = 2\phi\sqrt{1 - \frac{1}{4\phi^2}} = \sqrt{4\phi^2 - 1}$$

Equation 2.53

from which we get the numerical value:-

$$\tan\left(\frac{\pi}{5}\right) = 0.726542528005361$$

Equation 2.54

From Equation 2.53 we have:-

$$\left[\tan\left(\frac{\pi}{5}\right)\right]^2 = 4\phi^2\left(1 - \frac{1}{4\phi^2}\right) = 4\phi^2 - 1$$

Equation 2.55

The Pythagorean Theorem

The right-angled triangle whose acute angle is $\gamma = \pi/5$, has an Adjacent AJ that is unity; an Opposite OP that is $(4\phi^2-1)^{\frac{1}{2}}$; and a Hypotenuse HY of length 2ϕ.

These dimensions are summarised in the Figure 2.4

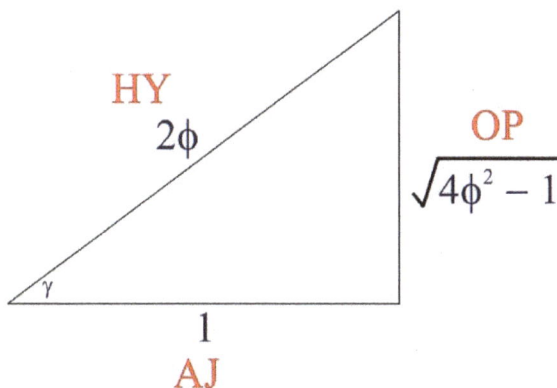

Figure 2.4
A Right-Angled Triangle with Acute Angle $\gamma = \pi/5$

It follows that the respective sides of the squares are:-

A   Adjacent, AJ, numerical value 1

B   Opposite, OP, value $\sqrt{4\phi^2 - 1}$, equal to
         0.726542528005361

C   Hypotenuse, HY, value 2ϕ, equal to
         1.23606797749979

whilst the squares of these sides are:-

1   Adjacent Squared, AJ^2, $1^2 = 1$

2   Opposite Squared, OP^2, $(4\phi^2 - 1)$
         $=0.527864045000421$

3   Hypotenuse Squared, HY^2, $4\phi^2 = 1.52786404500042$

Right-Triangle Area as One-Half its Rectangle, A_1

With regard to the simple right triangle of Figure 2.4 its First Area, A_1, is given by:-

$$A_1 = \frac{1}{2}.OP.AJ = \frac{1}{2}.\sqrt{4\phi^2 - 1}.1 = \frac{\sqrt{4\phi^2 - 1}}{2}$$
$$= 0.363271264002681$$

Equation 2.56

Obviously, this Area should be the same as the Triangle Areas calculated by other means, alternatives whose formulation may throw light upon relations between Pi and Phi.

Right Triangle Area by the Included Sine

This Area, A_2, is given by:-

$$A_2 = \frac{1}{2}.HY.AJ.\sin(\gamma) = \frac{1}{2}.2.\phi.1.\sin\left(\frac{\pi}{5}\right) = \phi.\sin\left(\frac{\pi}{5}\right)$$
$$= 0.363271264002681$$

Equation 2.57

Right Triangle Area by means of Heron's Theorem

Allow that the Semi-Perimeter of the Triangle, s, is given by:-

$$s = \frac{HY + AJ + OP}{2} = \frac{2\phi + 1 + \sqrt{4\phi^2 - 1}}{2}$$

$$= \frac{1}{2} + \phi + \frac{\sqrt{4\phi^2 - 1}}{2}$$

$$= 1.48130525275258$$

Equation 2.58

Then this Area, A$_3$, is given by:-

$$A_3 = \sqrt{s.(s - HY).(s - AJ).(s - OP)}$$
Equation 2.59

Now by substitution it transpires that:-

$$(s - HY) = \frac{1}{2} + \phi + \frac{\sqrt{4\phi^2 - 1}}{2} - 2\phi = \frac{1}{2} - \phi + \frac{\sqrt{4\phi^2 - 1}}{2}$$
Equation 2.60a

$$(s - AJ) = \frac{1}{2} + \phi + \frac{\sqrt{4\phi^2 - 1}}{2} - 1 = -\frac{1}{2} + \phi + \frac{\sqrt{4\phi^2 - 1}}{2}$$
Equation 2.60b

$$(s - OP) = \frac{1}{2} + \phi + \frac{\sqrt{4\phi^2 - 1}}{2} - \sqrt{4\phi^2 - 1} = \frac{1}{2} + \phi - \frac{\sqrt{4\phi^2 - 1}}{2}$$
Equation 2.60c

By de-constructing our expansion of Heron's Formula for the π/5 right-triangle we may ascertain that:-

$$A_3 = \sqrt{s(s - HY)} \times \sqrt{(s - AJ).(s - OP)} = H_1 + H_2$$

Equation 2.61

where:-

$$H_1 = \sqrt{\left(\frac{1}{2} + \phi + \frac{\sqrt{4\phi^2 - 1}}{2}\right)\left(\frac{1}{2} - \phi + \frac{\sqrt{4\phi^2 - 1}}{2}\right)}$$

Equation 2.62a

$$H_2 = \sqrt{\left(-\frac{1}{2} + \phi + \frac{\sqrt{4\phi^2 - 1}}{2}\right)\left(\frac{1}{2} + \phi - \frac{\sqrt{4\phi^2 - 1}}{2}\right)}$$

Equation 2.62b

It can be demonstrated that:-

$$A_3 = s(s - HY) = (s - AJ)(s - OP)$$

Equation 2.63

from which it follows that:-

$$A_3 = s(s - HY) = \left(\frac{1}{2} + \phi + \frac{\sqrt{4\phi^2 - 1}}{2}\right)\left(\frac{1}{2} - \phi + \frac{\sqrt{4\phi^2 - 1}}{2}\right)$$

Equation 2.64

Because ϕ is greater than zero we may write:-

$$A_3 = -\phi^2 + \frac{1}{4}.(4\phi^2 - 1) + \frac{\sqrt{4\phi^2 - 1}}{2} + \frac{1}{4}$$

Equation 2.65

so that condensation reveals that:-

$$A_3 = \frac{\sqrt{4\phi^2 - 1}}{2} = 0.363271264002681$$

Equation 2.66

and because:-

$$A_1 = A_2 = A_3$$
Equation 2.67

we may declare that:-

$$\sin\left(\frac{\pi}{5}\right) = \frac{OP}{HY} = \frac{\sqrt{4\phi^2 - 1}}{2\phi}$$
Equation 2.68

An Improvement of Series Convergence[2.7]

For immediate purposes assume that Iteration Limit m is locally seven.

This identity for the Cosine of $(\pi/5)$ is known:-

$$\cos\left(\frac{\pi}{5}\right) = \frac{1}{2\phi}$$
Equation 2.69

It remains to examine to what accuracy this cosine may be computed by some selected rapidly-convergent series.

Consider:-

$$\cos\left(\frac{\pi}{5}\right) = \sum_{k=0}^{m}(-1)^k \cdot \frac{\left(\frac{\pi}{5}\right)^{2k}}{(2k)!}$$

$$= \sum_{k=0}^{m}(-1)^k \cdot \frac{\left(\frac{\pi^{2k}}{5^{2k}}\right)}{(2k)!}$$

$$= \sum_{k=0}^{m}(-1)^k \cdot \frac{1}{(2k)!} \cdot \left(\frac{\pi^{2k}}{5^{2k}}\right)$$

Equation 2.70

All configurations yield fifteen-figure accuracy after seven iterations.

The Hyperbolic Cosine

The Hyperbolic Cosine of π/5 is given by:-

$$\cosh\left(\frac{\pi}{5}\right) = \frac{e^{\frac{\pi}{5}} + e^{-\frac{\pi}{5}}}{2} = 1.20397208933822$$

Equation 2.71

and the series of Equation 2.70 can be adapted to compute the hyperbolic cosine by the unsporting expedient of removing the alternator. Hence:-

$$\cosh\left(\frac{\pi}{5}\right) = \sum_{k=0}^{m} \frac{1}{(2k)!} \cdot \left(\frac{\pi^{2k}}{5^{2k}}\right)$$

Equation 2.72

This also delivers fifteen-figure accuracy when m = 7. It is notable that:-

$$\cosh(\phi) = 1.19714002066917$$

Equation 2.73

and accordingly the PSD discrepancy between $\cosh(\pi/5)$ and $\cosh(\phi)$ is:-

$$PSD\left[\cosh\left(\frac{\pi}{5}\right), \cosh(\phi)\right] = 0.567460718528908$$
Equation 2.74

The Laurent Series

Noting that:-

$$\Theta = \cosh\left[\cos\left(\frac{\pi}{5}\right)\right] = \cosh\left(\frac{1}{2\phi}\right) = \frac{e^{\frac{1}{2\theta}} + e^{-\frac{1}{2\theta}}}{2}$$
$$= 1.34549747270501$$
Equation 2.75

The first three terms of the Laurent Series for $Cosh(1/2\phi)$ (exclusive of unity) are:-

$$\Theta_{Laurent} = 1 + \frac{1}{8\phi^2} + \frac{1}{384\phi^4} + \frac{1}{46080\phi^4} = 1.34549288808723$$
Equation 2.76

for which the relevant Percentage Specific Defect is:

$$PSD(\Theta, \Theta_{Laurent}) = 0.000340737747452$$
Equation 2.77

Operationally, the general equation for the Laurent Series solution is:-

$$\Theta_{Laurent} = \sum_{k=0}^{m} \frac{4^{-k} \cdot \left(\frac{1}{\phi}\right)^{2k}}{(2k)!}$$
Equation 2.78

which delivers the same accuracy as Equation 2.76 when m = 3.

When m = 7 there is about twelve-figure accuracy as shown by:-

$$\Theta_{Laurent} = \sum_{k=0}^{7} \frac{4^{-k}.\left(\frac{1}{\phi}\right)^{2k}}{(2k)!} = 1.345497472705$$

Equation 2.79

and:-

$$PSD(\Theta, \Theta_{Laurent}) = 0.000000000000132$$

Equation 2.80

CHAPTER THREE
PHI

HER MODEST FOUR-THOUSAND-YEAR SPINSTERHOOD

First of all, like all females, Phi is a number with two complimentary aspects, often confounded, but always essential.

There is the Major Ratio of Phidias, Φ or PHI, which has a decimal value somewhere near to 1.61803398874989: And then there is the Minor Ratio of Phidias, φ of phi which is around 0.61803398874989. Sometimes, to augment the confusion, the Minor Ratio is called the Conjugate, but this terminology is best avoided unless special circumstances warrant it.

As you can see:-

$$\Phi = \phi + 1$$
Equation 3.1

By definition:-

$$\Phi = \frac{1 + \sqrt{5}}{2}$$
Equation 3.2

From which it follows that:-

$$\phi = \frac{\sqrt{5} - 1}{2}$$
Equation 3.3

By the way, when you consult literature you will often find Φ, or even φ, called a variety of confusing names such as the Golden Ratio, the Golden Section, the Golden Cut or any number of other things, both noble and base. And just to be extra confusing writers often use the Roman character F to represent the Fibonacci Number, a closely-related but not identical concept.

The Ratio of Phidias and the Division of Space

Though very many people associate the (Major and Minor) Ratio of Phidias with the arts, and especially with architecture, planning and the organisation of the composition of paintings, the concept has a basis in Nature and can be traced in the layout of the leaves and flowers of plants and also in the patterns of animal growth, such as the shell of the pearly nautilus (Nautilus pompilius) and various related extinct mollusks, such as the ammonites of the Mesozoic ages.

Modern work has shown that whilst the shell of Nautilus pompilius is indeed a logarithmic spiral in its profile, its growth ratio hovers around 1.4 rather than 1.618.

The Golden Rectangle[3.1]

The Golden Rectangle, like ϕ and Φ as algebraic constructs, exhibits a number of interesting recursive (self-similar) properties. It is a rectangle whose shorter side is unity and its longer The Major Phidian Ratio, otherwise known as the Golden Section, the Golden Mean or something, Φ, (PHI).

The Construction of the Golden Rectangle

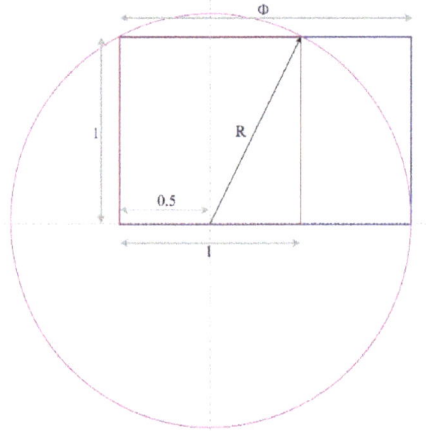

Figure 3.1
First Stages in the Generation of a Golden Rectangle

The first step to draft a Golden Rectangle is to construct a square. Of course you can do this in the dust of the ground, on a shard of stone, on paper, or upon your substratum of choice. Then you can bisect a convenient side of the square using the intersections of opposed arcs, or again by your method of choice. Placing your rope or compass at the point of square side bisection (0.5×side), open the Radius, R, to an opposing vertex of the square. Now draw a circle, or at least an adequate arc, until the circle intersects a line produced from the bisected square side. The square side plus the produced line is now of length Φ relative to the perpendicular square sides taken to be 1.

The right quadrilateral developed that has dimensions 1×Φ is a Golden Rectangle.

We shall discuss the real and alleged applications of the Golden Rectangle in due course. For the moment we shall discuss its theoretical aspects.

Figure 3.2 is a rough sketch diagram annotated with line, vertex and line labels discussed in mathematical derivations and demonstrations to follow.

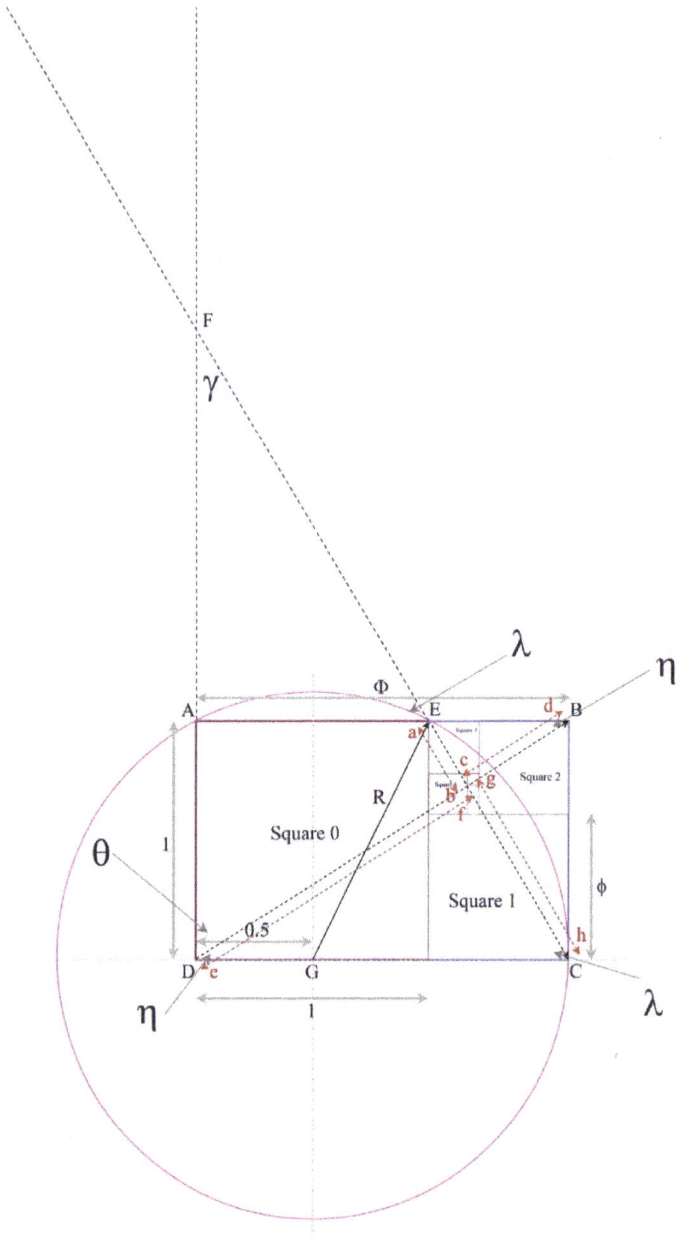

Figure 3.2
Definitional Diagram for the Golden Rectangle

ABCD is the Golden Rectangle. Note that when the square AD×AE is partitioned from the remainder of the Golden Rectangle, the remaining (right-hand) rectangle EB×BC is also Golden (in the sense of having its dimensions in the ratio Φ:1). This latter rectangle can also be decomposed into a Square 1, this time with dimensions $\phi \times \phi$, where ϕ (phi) is The Minor Ratio of Phidias. The residual rectangle EB×(1-ϕ) is also Golden and can be divided into a Square 2 of side $(1-\phi)^2$, and a remnant rectangle also golden containing Square 3, whose side is ϕ^3.

This program of sub-division can be continued *ad infinitum*, until the residual rectangles converge upon the Umbilicus of the Primary Rectangle ABCD at an offset point U (not shown in Figure 3.2).

This umbilical point U is at the intersection of the orthogonal diagonals DB and EC, which respectively divide into the segments ef+gd, and ab+gh.

The angles θ, η and λ relate in the following ways:-

$$\eta = \tan\left(\frac{1}{\Phi}\right)$$
Equation 3.4

$$\theta = \frac{\pi}{2} - \eta$$
Equation 3.5

$$\lambda = \tan\left(\frac{1}{\phi}\right)$$
Equation 3.6

$$\pi = \eta + \lambda + \frac{\pi}{2}$$
Equation 3.7

It obviously follows from Equation 3.7 that:-

$$\frac{\pi}{2} = \eta + \lambda$$
Equation 3.8

Figure 3.3 shows the central detail of the partitioning of the Golden Rectangle.

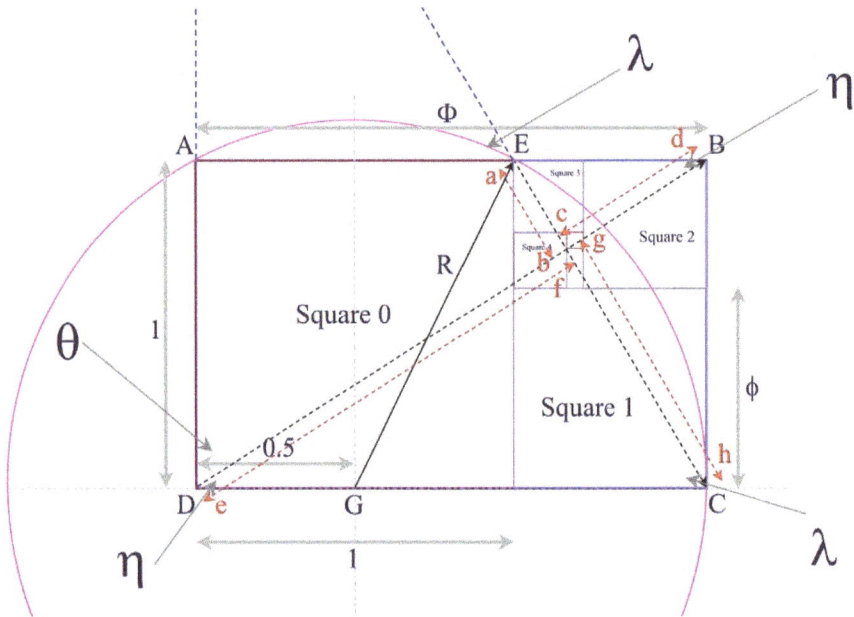

Figure 3.3
Definitional Diagram for the Golden Rectangle (detail)

The Recursion of Golden Rectangle Partitions

Φ and ϕ are of great importance in the treatment of mathematical recursion. Recursion is any self-similar process in which a procedure can be repeated in terms of itself to generate information. It is also is of great importance in the programming of machines.

I employed EXCEL® to generate slightly clearer and more accurate versions of Figure 3.3

Figure 3.4 resolves GE = R, the generative Radius, and also the first eight recursive squares of sub-division. U, the Umbilicus is explicitly indicated for the first time. It is now obvious that U lies somewhere in the rectangular region excluded by the little red Square 8

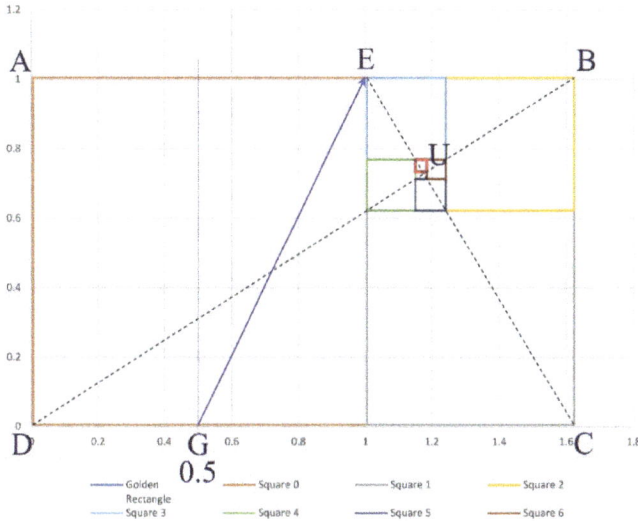

Figure 3.4
Clarification with the Convergence to the Umbilicus

Figure 3.5 is a development explicating the successive development of the areas of succeeding squares as:-

$$A_n = \phi^n$$
Equation 3.9

where A_n is the Area of Square n; ϕ is the Minor Ratio of Phidias; and n is the Square Serial Number starting with zero to denote the square in the initial constructed Golden Rectangle.

Figure 3.6 is a further development showing the Areas of the successive partition squares.

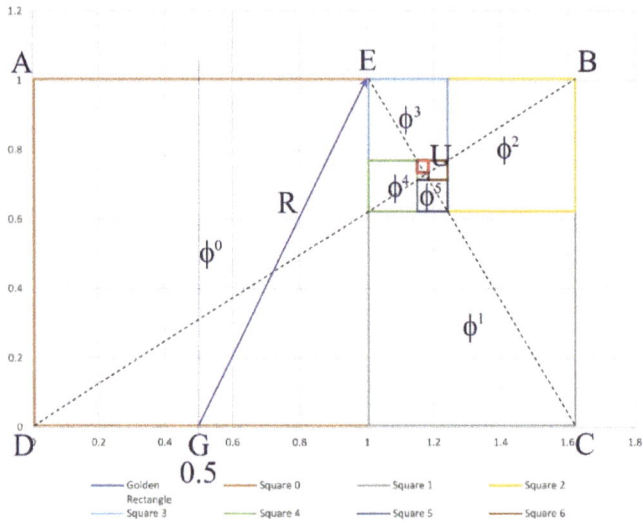

Figure 3.5
The Areas of Recursive Partition Squares
converging to the
Umbilicus

<u>The Umbilicus is Not Finitely Locatable</u>

The following are axiomatic to our system:-

$$\Phi = \frac{1 + \sqrt{5}}{2}$$

Equation 3.10

$$\phi = \Phi - 1$$
Equation 3.11

$$\Phi\phi = 1$$
Equation 3.12

Now the Length of the Line FD is the same as the First Equation Intercept a_{EC} as related to Φ and ϕ below:-

$$FD = a_{EC} = \frac{\Phi}{\phi} = 2 + \phi$$
Equation 3.13

and the Grade b_{EC} of EC is:-

$$b_{EC} = -\frac{1}{\phi}$$
Equation 3.14

Meanwhile, the Intercept, a_{DB} and Grade b_{DB} are given by:-

$$DB = a_{DB} = 0$$
Equation 3.15

$$b_{DB} = \frac{1}{\Phi} = \phi$$
Equation 3.16

Because EC and DB are therefore orthogonal the umbilicus can not be located by simultaneous linear equations, but might be (and is) approximated by iterative convergent procedures.

The umbilicus does not have a finite exact location because it is approximated by a sub-division process that itself has no natural termination. (You terminate the iteration by walking away from the exercise).

Further to demonstrate this issue, allow that a value of x is unity (or whatever other number you choose). The task is then to find x at U, the intersection of the two lines EC and DB.

The intersecting straight lines are defined by:-

$$y_{EC} = \frac{1}{\phi}(a_{EC} + b_{EC} \times x)$$
Equation 3.17a

and:-

$$y_{DB} = a_{DB} + b_{DB} \times x$$
Equation 3.17b

We also need to find the point where $y_{EC} = y_{DB}$. In other words where:-

$$y_{EC} = \frac{1}{\phi}(a_{EC} + b_{EC} \times x) = \frac{1}{\phi}\big(2 + \phi + (-(2 + \phi) \times x)\big) = 0$$
Equation 3.18a

and:-

$$y_{DB} = a_{DB} + b_{DB} \times x = 0 + \frac{1}{\Phi} \times x = 0.618033988749895 \dots$$
Equation 3.18b

Therefore:-

$$y_{EC} = \frac{1}{\phi}(0 \times x) = 0$$
Equation 3.19a

and:-

$$y_{DB} = \Phi\phi = 1$$
Equation 3.19b

Now Equations pairs 3.17 through 3.19 are all separately true but the final conclusion that $0 = 1$ is absurd, and so there is no finite umbilicus of the Golden Rectangle.

Or maybe you prefer the thought that there can be no finite outcome to a system involving irrational parameters.

Angular Relations

Now having to hand the Intercept of EC we may declare:-

$$\gamma = \tan^{-1}\left(\frac{\Phi}{a_{EC}}\right) = \tan^{-1}(\phi) = \tan^{-1}\left(\frac{\Phi}{2 + \phi}\right)$$

Equation 3.20

γ being the angle DFC in Figure 3.2
η is the angle BDC and θ is angle ADB:-

$$\eta = \tan^{-1}\left(\frac{1}{\Phi}\right)$$

Equation 3.21

$$\theta = \frac{\pi}{2} - \eta$$

Equation 3.22

κ, θ and η are the angles in Triangle FDU. They sum to π radians. Therefore:-

$$\kappa = \pi - \eta - \theta = \frac{\pi}{2}$$

Equation 3.23

Since angle DUF is right, the other three angles at the umbilicus are also right angles.
So:-

$$\lambda = \tan^{-1}\left(\frac{1}{\Phi - 1}\right) = \tan^{-1}\left(\frac{1}{\phi}\right)$$

Equation 3.24

and therefore the line FD measures:-

$$FD = \Phi \tan^{-1} \lambda$$

Equation 3.25

The Similar Triangles FCD and EBU

The Major Triangle FCD and the Minor Triangle EBU are similar because γ and η are equivalent.
Accordingly for adjacents:-

$$EB = \Phi - 1$$
Equation 3.26

$$FC = \sqrt{(2 + \phi)^2 + \Phi^2} = \sqrt{2\phi^2 + 6\phi + 5}$$
Equation 3.27

therefore for the diagonal component line cd:-

$$cd = FD\frac{EB}{FC} = (2 + \phi)\frac{\Phi - 1}{\sqrt{2\phi^2 + 6\phi + 5}} = \frac{\phi(2 + \phi)}{\sqrt{2\phi^2 + 6\phi + 5}}$$
Equation 3.28

And for opposites:-

$$DC = \Phi$$
Equation 3.29

$$ab = DC\frac{EB}{FC} = \frac{\Phi(\Phi - 1)}{\sqrt{2\phi^2 + 6\phi + 5}}$$
Equation 3.30

If you desire a confirmatory integrity check a possibility is:-

$$\frac{EB}{FC} = \frac{ab}{DC} = \frac{cd}{FD} \approx 0.200811415886227$$
Equation 3.31

By the rule of opposite angles, Triangles EBU and DCU have two angles in common, η and opposite umbilical right-angles.
Therefore the triangles EBU and DCU are similar.
For DCU:-

$$DC = \Phi = \phi + 1 \approx 1.61803398874989 \ldots$$
$$\textbf{Equation 3.32}$$

whilst:-

$$EB = \Phi - 1 \approx 0.618033988749895 \ldots$$
$$\textbf{Equation 3.33}$$

Accordingly for adjacents:-

$$ef = cd\,\frac{DC}{EB} = \frac{\phi(2 + \phi)}{\sqrt{2\phi^2 + 6\phi + 5}} \times \frac{\phi + 1}{\Phi - 1}$$
$$\textbf{Equation 3.34}$$

And for opposites:-

$$gh = ab\,\frac{DC}{EB} = \Phi\,\frac{\Phi - 1}{\sqrt{2\phi^2 + 6\phi + 5}} \times \frac{\phi + 1}{\Phi - 1} = \frac{\Phi(\phi + 1)}{\sqrt{2\phi^2 + 6\phi + 5}}$$
$$\textbf{Equation 3.35}$$

If you desire a confirmatory integrity check a possibility is:-

$$\frac{DC}{EB} = \frac{gh}{ab} = \frac{ef}{cd} = \Phi + 1 \approx 2.618033988749895 \ldots$$
$$\textbf{Equation 3.36}$$

The Cartesian Co-ordinates of the Umbilicus

It is possible to derive cartesian co-ordinates for the umbilicus in terms of other irrational numbers though not by finite methods such as simultaneous linear algebraic equations. This is so even though it was shown long ago that the Phidian ratios have algebraic polynomial roots and are therefore not transcendental.

Allow that x_U is the X-coordinate of the Golden Rectangle Umbilicus and that y_U is the Y-coordinate of the Golden Rectangle Umbilicus.

Then note that:-

$$x_U = ef.\cos^{-1}\left(\tan^{-1}\left(\frac{1}{\Phi}\right)\right) = ef.\frac{1}{\sqrt{1 + \left(\frac{1}{\Phi}\right)^2}}$$

Equation 3.37

so:-

$$x_U = \frac{\phi(2 + \phi)}{\sqrt{2\phi^2 + 6\phi + 5}} \times \frac{\phi + 1}{\Phi - 1} \times \frac{1}{\sqrt{1 + \left(\frac{1}{\Phi}\right)^2}}$$

Equation 3.38

Correspondingly:-

$$y_U = ef.\sin^{-1}\left[\tan^{-1}\left(\frac{1}{\Phi}\right)\right] = ef.\frac{\frac{1}{\Phi}}{\sqrt{1 + \left(\frac{1}{\Phi}\right)^2}} = ef.\frac{1}{\Phi\sqrt{1 + \left(\frac{1}{\Phi}\right)^2}}$$

Equation 3.39

so:-

$$y_U = \frac{\phi(2 + \phi)}{\sqrt{2\phi^2 + 6\phi + 5}} \times \frac{\phi + 1}{\Phi - 1} \times \frac{1}{\Phi\sqrt{1 + \left(\frac{1}{\Phi}\right)^2}}$$

Equation 3.40

Simplifications then yield:-

$$x_U = (\phi + 1)\frac{2 + \phi}{3 + \phi}$$

Equation 3.41

and:-

$$y_U = \frac{2 + \phi}{3 + \phi}$$

Equation 3.42

from which it is manifest that:-

$$\frac{x_U}{y_U} = \Phi$$

Equation 3.43

In numerical terms:-

$$x_U = \Phi\frac{2+\phi}{3+\phi} \approx 1.17082039324994\ldots$$

Equation 3.44

$$y_U = \frac{1+\Phi}{2+\Phi} \approx 0.723606797749979\ldots$$

Equation 3.45

<u>Partitioned Areas of the Golden Rectangle</u>

The Radius R = EG

The Generative Radius R is numerically:-

$$R = \sqrt{(0.5)^2 + 1^2} = 1.11803398874989$$

Equation 3.46

Technically this "equation" is of course an irrational approximation but for normal scientific and engineering purposes it will do.

The Quadrilateral AEGD

The Area of this figure that covers most of Square 0 is numerically:-

$$A_{AEGD} = 1\frac{1}{2} + 1\frac{1}{2}\frac{1}{2} = 0.75$$

Equation 3.47

The Triangle EGC

The Area of Triangle EGC is:-

$$A_{EGC} = \frac{1}{2}(ab + gh)R\sin(\lambda) = \frac{2\phi + 1}{4}$$
Equation 3.48

The Triangle EBC

The Area of Triangle EBC is:-

$$A_{EBC} = \frac{\phi \times 1}{2} = \frac{\phi}{2}$$
Equation 3.49

Summation

Since the Area of the Golden Rectangle ABCD is given by:-

$$A_{ABCD} = \Phi \times 1 = \Phi$$
Equation 3.50

It is readily confirmed that:-

$$A_{ABCD} = A_{AEGD} + A_{EGC} + A_{EBC}$$

$$= \frac{3}{4} + \frac{2\phi + 1}{4} + \frac{2\phi}{4} = \frac{4\phi + 4}{4} = \phi + 1 = \Phi$$
Equation 3.51

The Sum and Product of the Minor Ratio Power Squares

The space-filling Sum of the Minor Phidian Ratio Power Squares is defined to be:-

$$S_P = \sum_{k=0}^{\infty} \phi^k$$
Equation 3.52

Clearly, infinity is not attainable in practice so that in numerical experiments I adopted either m = 128 or m = 256 as a sufficient surrogate.

I intuitively anticipated that S_P would approach the Golden Rectangle Area, Φ. Consult Figure 3.5 to see how the ever-decreasing squares seem to fill the Golden Rectangle.

I set m = 128 and the actual S_P was found to be:-

$$S_P = \sum_{k=0}^{m} \phi^k = \Phi + 1 \approx 2.6180339887499 \ldots$$

Equation 3.53

So much for my intellectual intuition.

Meanwhile the associated iterated Power Product was more understandably:-

$$P_P = \prod_{k=0}^{m} \phi^k = 0$$

Equation 3.54

Series Renditions

Allow that, with m = 256, and understanding that the subscript labels i = 1,5..m/4 and j = 3, 7..m/4, the umbilical sums $y_{U\Sigma A}$ and $y_{U\Sigma B}$ are defined as follows:-

$$y_{U\Sigma A} = \sum_{i}^{m} (\phi^i + \phi^{i+1}) \approx 1.17082039324989$$

Equation 3.55

$$y_{U\Sigma B} = \sum_{j}^{m} (-\phi^j - \phi^{j+1}) \approx -0.447213595499939$$

Equation 3.56

then:-

$$y_{U\Sigma} = y_{U\Sigma A} + y_{U\Sigma B}$$
Equation 3.57

$$y_{U\Sigma} = \sum_{i}^{m}\left(\phi^i + \phi^{i+1}\right) + \sum_{j}^{m}\left(-\phi^j - \phi^{j+1}\right)$$
Equation 3.58

$$y_{U\Sigma} = \sum_{i}^{m}\left(\phi^i + \phi^{i+1}\right) - \sum_{j}^{m}\left(\phi^j + \phi^{j+1}\right)$$
Equation 3.59

The Diagonals of the Golden Rectangle (In Line Operations)

We wish to study the systems:-

$$ef^2 - cd^2 \sim ab^2 - gh^2$$
System 3.1

$$ef^2 + cd^2 \sim ab^2 + gh^2$$
System 3.2

where the symbol $\sim$ indicates "postulated relationship" not necessarily equality.

Squares of Diagonal Line Components

From the equations for ef, cd, ab and gh we may establish that:-

$$ef^2 = \frac{\phi^4 + 6\phi^3 + 13\phi^2 + 12\phi + 4}{2\phi^2 + 6\phi + 5}$$
Equation 3.60a

$$cd^2 = \frac{\phi^4 + 4\phi^3 + 4\phi^2}{2\phi^2 + 6\phi + 5}$$

Equation 3.60b

$$ab^2 = \frac{\phi^4 + 2\phi^3 + \phi^2}{2\phi^2 + 6\phi + 5}$$

Equation 3.60c

$$gh^2 = \frac{\phi^4 + 4\phi^3 + 6\phi^2 + 4\phi + 1}{2\phi^2 + 6\phi + 5}$$

Equation 3.60d

Also:-

$$gh^2 = \frac{\phi^4 + 4\phi^3 + 6\phi^2 + 4\phi + 1}{2\phi^2 + 6\phi + 5} = \frac{2 + \phi}{3 + \phi} = \frac{\Phi + 1}{\Phi + 2}$$

Equation 3.61

Moving forward to study the expression ef²-cd² we may write:-

$$ef^2 - cd^2 = \frac{\phi^4 + 6\phi^3 + 13\phi^2 + 12\phi + 4}{2\phi^2 + 6\phi + 5} - \frac{\phi^4 + 4\phi^3 + 4\phi^2}{2\phi^2 + 6\phi + 5} = \Phi$$

Equation 3.62

$$ef^2 - cd^2 = \frac{2\phi^3 + 9\phi^2 + 12\phi + 4}{2\phi^2 + 6\phi + 5} = \Phi$$

Equation 3.63

and:-

$$gh^2 - ab^2 = \frac{\phi^4 + 4\phi^3 + 6\phi^2 + 4\phi + 1}{2\phi^2 + 6\phi + 5} - \frac{\phi^4 + 2\phi^3 + \phi^2}{2\phi^2 + 6\phi + 5} = \phi$$

Equation 3.64

$$gh^2 - ab^2 = \frac{2\phi^3 + 5\phi^2 + 4\phi + 1}{2\phi^2 + 6\phi + 5} = \phi$$
Equation 3.65

because $\Phi = (\phi^3+3)/2$ and $\phi = 0.618033988749895$
Therefore:-

$$(ef^2 - cd^2) - (gh^2 - ab^2) = \frac{\phi^3 - 2\phi + 3}{2} = 1$$
Equation 3.66

and:-

$$\frac{(2\phi + 1)(2\phi + 3)}{2\left(\phi + \frac{3}{2}\right)^2 + \frac{1}{2}} = \frac{4\phi^2 + 8\phi + 3}{2\phi^2 + 6\phi + 5} = 1$$
Equation 3.67

The addition give us:-

$$(ef^2 - cd^2) + (gh^2 - ab^2) = \phi^3 + 2$$
Equation 3.68

Squares of Diagonal Line Components (Cross Operations)

   In this exercise we look at the relations of the squares and sums (or differences) involving the opposites and adjacents of the similar triangles DUC and BUE.
   I.e.:-

$$(ef^2 - gh^2) = X_{DUC}$$
Equation 3.69

and:-

$$(cd^2 - ab^2) = X_{BUE}$$
Equation 3.70

So:-

$$X_{DUC} = ef^2 - gh^2 = \frac{2\phi^3 + 7\phi^2 + 8\phi + 3}{2\phi^2 + 6\phi + 5}$$
Equation 3.71

$$X_{DUC} = ef^2 - gh^2 = \frac{2(\phi + 1)^3 + (\phi + 1)^2}{2\phi^2 + 6\phi + 5} = \frac{2\Phi^3 + \Phi^2}{2\Phi^2 + 2\Phi + 1}$$
Equation 3.72

and:-

$$X_{BUE} = cd^2 - ab^2 = \frac{2\phi^3 + 3\phi^2}{2\phi^2 + 6\phi + 5} = \frac{\Phi}{2\Phi^2 + 2\Phi + 1}$$
Equation 3.73

So we may complete this study of cross-lengths in similar triangles by observing that:-

$$X_{DUC} - X_{BUE} = (ef^2 - gh^2) - (cd^2 - ab^2) = 1$$
Equation 3.74

and:-

$$X_{DUC} + X_{BUE} = (ef^2 - gh^2) + (cd^2 - ab^2) = \frac{2\Phi^3 + \Phi^2 + \Phi}{2\Phi^2 + 2\Phi + 1}$$
Equation 3.75

If convenient this last expression can be processed as:-

$$X_{DUC} + X_{BUE} = \frac{4\left(\Phi + \frac{1}{6}\right)^3 + \frac{10}{6}\left(\Phi + \frac{1}{6}\right) - \left(\frac{2}{3}\right)^3}{4\left(\Phi + \frac{1}{2}\right)^2 + 1}$$
Equation 3.76

The Golden Spiral

The Golden Spiral is a special case of a Logarithmic Spiral whose general polar equation may be specified as:-

$$r = ae^{b\theta}$$
Equation 3.77

where r is a local Spiral Radius which may usefully be centered at the Umbilicus of the Golden Rectangle (x_U, y_U). a is some Constant and b is another Constant, which in practical terms is negative. θ (theta) is some Lapse Angle centered upon the umbilicus. Finally, e is the Napierian Base, sometimes called Euler's Number or other things, a transcendental number with a value near to 2.718281828459045

Various formats of Golden Spiral are found in the literature, conditioned by various rigid rotations, translations and scalings in order to demonstrate various principles.

The Golden Spiral is supposed to have a number of important significances in the arts and sciences and we shall organise ours very nearly to intercept the diagonal corners of our component squares of sides ϕ^0, ϕ^1, ϕ^2, ϕ^3, ϕ^4, ϕ^5,... as seen in the golden rectangle of Figure 3.5.

Please note that a properly-calculated golden spiral does *not* intercept the squares' diagonals *exactly*. Often, artists and teachers and others connect the squares' corners with circular arcs: This gives a very good visual approximation of a logarithmic spiral which is already an approximation of an unhaveable transcendental figure.

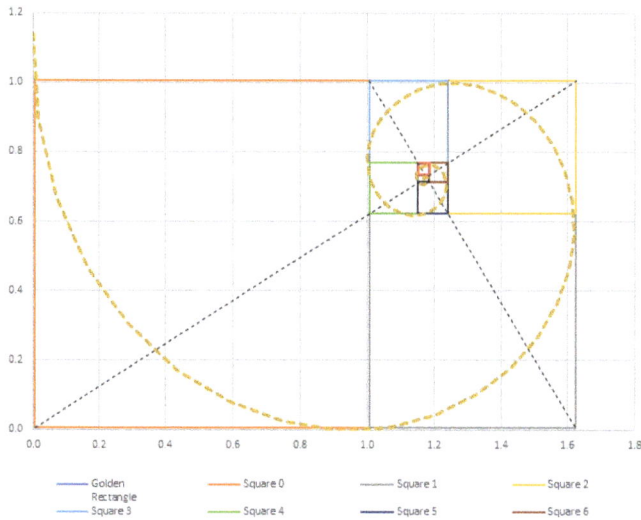

Figure 3.6
The Golden Spiral with
The Areas of Recursive Partition Squares
converging to the
Umbilicus

As you can see the spiral misses the vertices of the squares at several points. This is the issue of my poor computation and draftsmanship, but serves to remind us of the fact that a golden spiral is an imprecise animal!

Generation of the Golden Spiral

For the n = 128 points, labelled i = 0 .. n the spiral of Figure 3.6 was defined by:-

$$a = 1.94649798$$
Equation 3.78

$$b = -0.30634896$$
Equation 3.79

and:-

$$\theta_{start} = -0.1097 + \pi - \tan^{-1}\left(\frac{1 - y_U}{x_U}\right)$$

Equation 3.80

The Polar Co-ordinates θ_i and r_i are then defined by:-

$$\theta_i = \theta_{start} + v.\,2\pi\frac{i}{n}$$

Equation 3.81

and:-

$$r_i = a.\,e^{b.\theta_i}$$

Equation 3.82

v is the Number of Volutes Required, in our instance four.

To render these points in Cartesian terms (for plotting in EXCEL® etcetera) we may use:-

$$x_i = x_U + x_{dil}.\,r_i \cos(\theta_i)$$

Equation 3.83

$$y_i = y_U + y_{dil}.\,r_i \sin(\theta_i)$$

Equation 3.84

The Golden Angle

The Golden Angle is that portion of a circle (which of course has a complete polar sweep of 2π radians) which divides the circle in the Ratio:-

$$\frac{2\pi(1 - \phi)}{2\pi} = \frac{\beta}{2\pi} = 1 - \phi = \frac{1}{\Phi^2} = \frac{\alpha}{\beta} = f = 0.381966011250105$$

Equation 3.85

where α and β are Angles; f is the Fraction of the Circle Sweep accounted for by The Golden Angle, γ; and ϕ and Φ are respectively the Minor and the Major Ratios of Phidias.

It follows that γ, The Golden Angle is given by:-

$$\gamma = 2\pi f = 2\pi \left(1 - \frac{1}{\Phi}\right) = 2\pi(1 - \phi) = \frac{2\pi}{\Phi^2} = \pi\left(3 - \sqrt{5}\right)$$

Equation 3.86

The $\text{Cos}(\gamma)$ and $\text{Sin}(\gamma)$ are both transcendental, and γ itself has no exact value. The Golden Angle, γ, is approximately 2.39996322972865 radians or 137.507764050038 degrees.

As a corollary of these facts it is notable that Start Angle Constant, κ, is very roughly 3.735 *for sinistral generation of the arc from the top of the circle* in the equation:-

$$\psi_0 = \kappa\pi$$

Equation 3.87

where ψ_0 is the Start Angle, and the generative equation is taken to be:-

$$\psi_j = \psi_0 + j.\frac{\beta}{n}$$

Equation 3.88

for elements j = 1...m

The Golden Angle is represented by the red outer arc against the blue inner circle in Figure 3.7:-

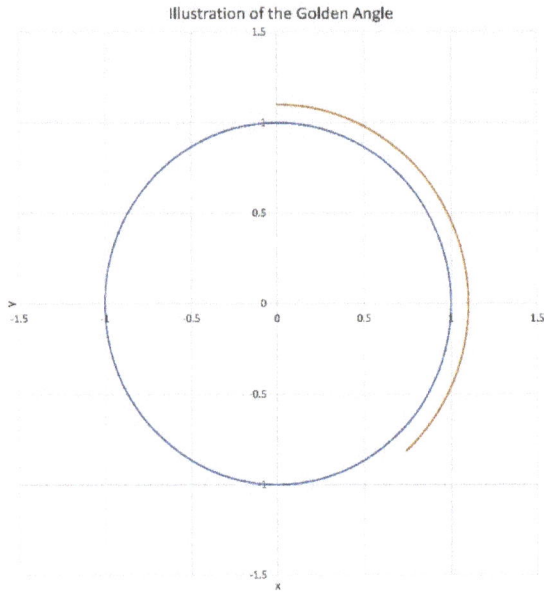

Figure 3.7
The Golden Angle as a Portion of the Circle

The Golden Rhombus

The Golden Rhombus is a diamond or lozenge-shaped quadrilateral whose Major x-Axis, a, is z = Φ/2; and Minor y-Axis, b, is ½.

Accordingly it can be inscribed within the Golden Rectangle for whose area it accounts for a half. (I.e. The Area of the Golden Rhombus is z = Φ/2).

The geometry is clarified in Figure 3.8:-

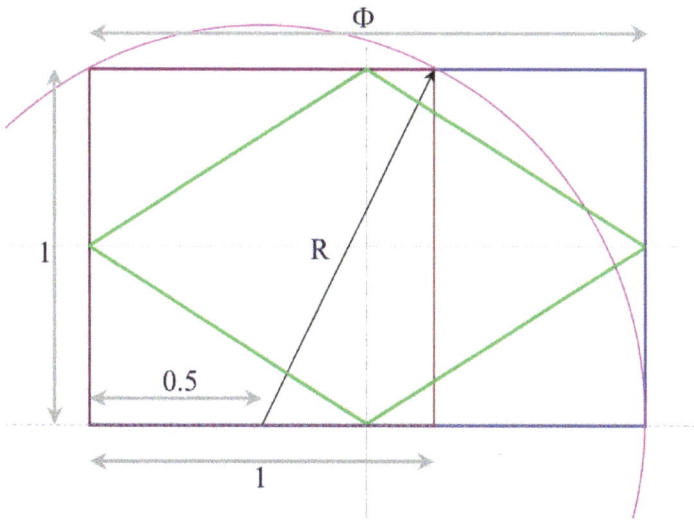

Figure 3.8
The Golden Rhombus

It is now plain to see that the Circumference, C_R, of the Rhombus is:-

$$C_R = 4\sqrt{a^2 + b^2} = 4\sqrt{\left(\frac{\Phi}{2}\right)^2 + \left(\frac{1}{2}\right)^2} = 4\sqrt{\left(\frac{\Phi^2}{4}\right) + \left(\frac{1}{4}\right)}$$

Equation 3.89

from which it follows that:-

$$C_R = 4\sqrt{\frac{\Phi^2 + 1}{4}} = 2\sqrt{\Phi^2 + 1}$$

Equation 3.90

It is also evident that any closed convex hull figure that is produced to inscribe the Golden Rectangle, but also to circumscribe the Golden Rhombus must have both a Circumference of intermediate value; and simultaneously an Area of intermediate value.

Let A_X be the area of this so-far-undefined shape, and C_X be its Circumference. A_R and C_R are respectively the Area and Circumference of the Golden Rhombus and A_G and C_G respectively the Area and the Circumference of the Golden Rectangle.

Then:-

$$A_R < A_X < A_G$$
Inequality 3.1

and:-

$$C_R < C_X < C_G$$
Inequality 3.2

or by substitution:-

$$\frac{\Phi}{2} < A_X < \Phi$$
Inequality 3.3

and:-

$$2\sqrt{\Phi^2 + 1} < C_X < 2(\Phi + 1)$$
Inequality 3.4

The Golden Ellipse[3.2]

Figure 3.9 shows the Golden Ellipse set within the Golden Rectangle and outwith the Golden Rhombus.

The Area of the Ellipse

In general, the Area of an Ellipse, A_{elli}, is given by:-

$$A_{elli} = \pi a b$$
Equation 3.91

where a and by are respectively the Major and Minor Semi-Axes.

In terms of the Golden Ellipse Area, A_E, this can be stated as:-

$$A_E = \pi \frac{\Phi}{2} \frac{1}{2} = \frac{\pi \Phi}{4}$$
Equation 3.92

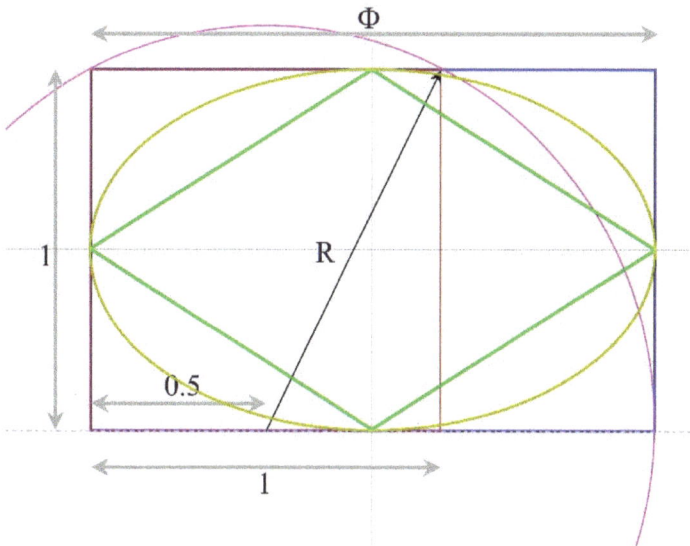

Figure 3.9
The Golden Rhombus

Areal Relations

The Areas of the Golden Rhombus, Golden Ellipse and Golden Rectangle are respectively styled A_R, A_E and A_G.

Figure 3.9 confirms that:-

$$A_R < A_E < A_G$$
Inequality 3.5a

Since:-

$$A_R = \frac{\Phi}{2} \approx 0.809016994374947$$
Equation 3.93a

$$A_E = \frac{\pi\Phi}{4} \approx 1.27080092307881$$
Equation 3.93b

$$A_G = \Phi \approx 1.61803398874989$$
Equation 3.93c

We may write:-

$$\frac{\Phi}{2} < \frac{\pi\Phi}{4} < \Phi$$
Inequality 3.5b

Accordingly it follows that:-

$$\frac{A_E}{A_R} = \frac{\pi\Phi}{4} \times \frac{2}{\Phi} = \frac{\pi}{2}$$
Equation 3.94

and:-

$$\frac{A_G}{A_E} = \frac{\Phi}{\frac{\pi\Phi}{4}} = \frac{4\Phi}{\pi\Phi} = \frac{4}{\pi}$$
Equation 3.95

The Circumference of the Ellipse: Definitions

First of all, note this special definition of the Conjugate of the Major Ratio of Phidias, ψ:-

$$\psi = \frac{1 - \sqrt{5}}{2} = -\phi \approx -0.618033988749895$$

Equation 3.96

This is the negative of many "conjugates" you will find in the literature, which are tantamount to the Minor Ratio, ϕ.

Until further notice we shall work with this Conjugate ψ.

Among other local introductions we may view:-

$$a \equiv z = \frac{\Phi}{2}$$

Equation 3.97

$$b = \frac{1}{2}$$

Equation 3.98

$$c = \sqrt{a^2 - b^2} = \sqrt{\frac{\Phi^2}{4} - \frac{1}{4}} = \frac{1}{2}\sqrt{\Phi^2 - 1}$$

Equation 3.99

with:-

$$h = \frac{(a - b)^2}{(a + b)^2}$$

Equation 3.100

The ellipse parameter Ellipticity, ε, is given by:-

$$\varepsilon = \frac{c}{a} = \sqrt{1 - \frac{b^2}{a^2}}$$

Equation 3.101

The Gamma Function, $\Gamma(x)$, is a species of false factorial generalised for the case of Real Numbers (Real Numbers include [decimal] fractions as well as integers). Among its many employments is a role in the computation of the circumference of any ellipse, a length with no elementary solution.

The Gamma Function, $\Gamma(x)$, is given by the integral:-

$$\Gamma(x) = \int_0^\infty t^{x-1}.e^{-t}.dt$$

Equation 3.102

Obviously, infinity ∞ is inaccessible, but Mathcad® will yield fourteen figure accuracy in the estimation of $\Gamma(x)$ if the upper bound of the integral is taken to be 48.

Indeed, a possible value for our purposes would be:-

$$\Gamma(8.5) = \int_0^{48} t^{x-1}.e^{-t}.dt = 1403.4407293478$$

Equation 3.103

There exist convenient special forms of the Gamma Function for Odd and for Even Integers, forms that involve true factorials only and may conveniently be summarised by the functions NODD(k) and NEVEN(k) respectively:-

$$NODD(k) = \frac{(2k-1)!}{2^{k-1}.(k-1)!}$$

Equation 3.104

and:-

$$NEVEN(k) = 2^k.k!$$

Equation 3.105

The Circumference of the Ellipse: Definition by Elliptic Integral

The Circumference of the Ellipse, C_E, is given by:-

$$C_E = 4a.E(\varepsilon)$$
Equation 3.106

where $E(\varepsilon)$ is the Complete Elliptic Integral of the Second Kind.

ε is the Eccentricity of the ellipse defined as:-

$$\varepsilon = \sqrt{1 - \frac{b^2}{a^2}}$$
Equation 3.107

and the Complete Elliptic Integral of the Second Kind is:-

$$E(\varepsilon) = \int_0^{\pi/2} \sqrt{1 - \varepsilon^2.\sin(\theta)^2}.d\theta$$
Equation 3.108

Allow that the Circumference of an Ellipse is C, then the Equation of the Circumference of an Ellipse in terms of the Gamma Function approach may be written:-

$$C = \pi(a+b)\left\{1 + \sum_{n=1}^{m}\left[\frac{NODD(m)}{2^n.n!}\right]^2 . \frac{h^n}{(2n-1)^2}\right\}$$
$$= 4.16989459467391$$
Equation 3.109

or:-

$$C = \pi(a+b)\left\{1 + \sum_{n=1}^{m}\left[\frac{1}{2^n.n!} \times \frac{(2n-1)!}{2^{n-1}.(n-1)!}\right]^2 . \frac{h^n}{(2n-1)^2}\right\}$$
$$= 4.16989459467391$$

Equation 3.110

As you have probably anticipated, this and other Ellipse Circumference equations may be re-formulated and indeed simplified in a diversity of ways. Your choice of formula will be informed by economy and the minimisation of numerical error.

I found that in this elaboration m was best as nine.

<u>The Circumference of the Ellipse: Substitutive Simplifications</u>

We have already seen how c, a hypotenuse parameter, may severally be simplified and it follows from that that the Ellipticity ε may be quoted as:-

$$\varepsilon = \frac{c}{a} = \frac{\sqrt{\Phi^2 - 1}}{\Phi} = \sqrt{1 - \frac{1}{\Phi^2}} = 0.63600924757034$$

Equation 3.111

Reverting to m = 8 I then simplified the Elliptic Integral for Phidian parameters as:-

$$E(\varepsilon) = \int_0^{\pi/2} \sqrt{1 - \left(\frac{\sqrt{\Phi^2 - 1}}{\Phi}\right)^2 (\sin(\theta))^2} \, . \, d\theta = 1.28856829450647$$

Equation 3.112

Hence:-

$$C = 4 . \frac{\Phi}{2} . E(\varepsilon) = 2\Phi E(\varepsilon) = 4.1698945946739$$

Equation 3.113

And with regard to the component h:-

$$h = \frac{(a-b)^2}{(a+b)^2} = \frac{\left(\frac{\Phi}{2} - \frac{1}{2}\right)^2}{\left(\frac{\Phi}{2} + \frac{1}{2}\right)^2} = \frac{\left(\frac{\Phi-1}{2}\right)^2}{\left(\frac{\Phi+1}{2}\right)^2} = \frac{\Phi^2 - 2\Phi + 1}{\Phi^2 + 2\Phi + 1}$$

$$= \frac{(\Phi-1)(\Phi-1)}{(\Phi+1)(\Phi+1)}$$

Equation 3.114

The Circumference of the Ellipse: The Ivory-Bessel Series Formula

Employing the above substitutes the Circumference of the Golden Ellipse may be given as:-

$$C = \pi(a+b)\left\{1 + \sum_{n=1}^{m}\left[\frac{\frac{(2n-1)!}{2^{n-1}.(n-1)!}}{2^n.n!}\right]^2 \cdot \frac{\left(\frac{(\Phi-1)(\Phi-1)}{(\Phi+1)(\Phi+1)}\right)^n}{(2n-1)^2}\right\}$$

$$= 4.169894594467391$$

Equation 3.115

Now it is manifest that:-

$$\frac{\frac{(2n-1)!}{2^{n-1}.(n-1)!}}{2^n.n!} = \frac{(2n-1)!}{2^{n-1}.(n-1)!} \times \frac{1}{2^n.n!} = \frac{(2n-1)!}{2^n.n!.2^{n-1}.(n-1)!}$$

Equation 3.116

which by expansion and cancellation may be rendered:-

$$\frac{\prod_{k=n}^{2n-1} k}{2^{2n-1}.(n)!} = \frac{\prod_{k=n+1}^{2n-1} k}{2^{2n-1}.(n-1)!} = 0.2734375$$

Equation 3.117

A second simplification is given by:-

$$\left(\frac{1}{2^{2n-1}}\right)^2 = \left(\frac{1}{2^{2n-1}}\right)\left(\frac{1}{2^{2n-1}}\right) = \frac{1}{2^{2(2n-1)}} = 0.00006103515625$$

Equation 3.118

and a third by:-

$$\left(\prod_{k=n}^{2n-1} k\right)\prod_{k=n}^{2n-1} k = \prod_{k=n}^{2n-1} k^2 = \left(\prod_{k=n}^{2n-1} k\right)^2 = 705600$$

Equation 3.119

or as an approximation:-

$$\frac{2^{4n-2}.\left[\Gamma\left(n+\frac{1}{2}\right)\right]^2}{\pi} = 705600.003435333$$

Equation 3.120

These various simplifications and re-statements enable us to re-quote the Ivory-Bessel Series for Φ as:-

$$C = \pi\left(\frac{\Phi+1}{2}\right)\left\{1 + \sum_{n=1}^{m}\left[\frac{(2n-1)!}{2^n.n!.2^{n-1}.(n-1)!}\right]^2 . \frac{\left(\frac{(\Phi-1)(\Phi-1)}{(\Phi+1)(\Phi+1)}\right)^n}{(2n-1)^2}\right\}$$

$$= 4.16989459467391$$

Equation 3.121

or:-

$$C = \pi\left(\frac{\Phi+1}{2}\right)\left\{1 + \sum_{n=1}^{m}\left[\frac{\prod_{k=n}^{2n-1} k}{2^{2n-1}.(n)!}\right]^2 . \frac{\left(\frac{(\Phi-1)(\Phi-1)}{(\Phi+1)(\Phi+1)}\right)^n}{(2n-1)^2}\right\}$$

$$= 4.16989459467391$$

Equation 3.122

or:-

$$C = \pi\left(\frac{\Phi + 1}{2}\right)\left\{1 + \sum_{n=1}^{m}\left(\frac{1}{2^{2\times(2n-1)}}\right)\left[\frac{\prod_{k=n}^{2n-1} k^2}{(n)!^2}\right] \cdot \frac{\left(\frac{(\Phi - 1)(\Phi - 1)}{(\Phi + 1)(\Phi + 1)}\right)^{n}}{(2n - 1)^2}\right\}$$

$$= 4.169894594467391$$

Equation 3.123

Note that all three variants of C have the same "exact" numerical value: Substitution of the integral for $\Gamma(n+\frac{1}{2})$ would yield 4.169894594(44833)

Further replacements allow us to write:-

$$C = \pi\left(\frac{\Phi + 1}{2}\right)\left\{1 + \frac{1}{\pi}\sum_{n=1}^{m}\left[\frac{\left(\sqrt{\pi}\frac{(2n)!}{4^n . n!}\right)^2}{(n!)^2}\right] \cdot \frac{\left[\frac{\phi^2}{(\Phi + 2)^2}\right]^{n}}{(2n - 1)^2}\right\}$$

$$= 4.16989459467391$$

Equation 3.124

or:-

$$C = \pi\left(\frac{\Phi + 1}{2}\right)\left\{1 + \frac{1}{\pi}\sum_{n=1}^{m}\left[\frac{\left(\sqrt{\pi}.(2n)!\right)^2}{(4^n . n!)^2.(n!)^2}\right] \cdot \frac{\left[\frac{\phi^2}{(\Phi + 2)^2}\right]^{n}}{(2n - 1)^2}\right\}$$

$$= 4.16989459467391$$

Equation 3.125

Equation 3.125 eventuates in the variants:-

$$C = \pi\left(\frac{\Phi + 1}{2}\right)\left\{1 + \frac{1}{\pi}\sum_{n=1}^{m}\left[\frac{\pi.(2n)!^2}{4^{2n}.(n!)^2.(n!)^2}\right] \cdot \frac{\left[\frac{\phi^2}{(\Phi + 2)^2}\right]^{n}}{(2n - 1)^2}\right\}$$

$$= 4.16989459467391$$

Equation 3.126

and:-

$$C = \pi\left(\frac{\Phi + 1}{2}\right)\left\{1 + \sum_{n=1}^{m}\left[\frac{1}{4^{2n}}\right] \cdot \left[\frac{\prod_{k=n+1}^{2n} k}{n!}\right]^2 \cdot \frac{\left[\frac{\phi^2}{(\Phi + 2)^2}\right]^n}{(2n - 1)^2}\right\}$$

$$= 4.16989459467391$$

Equation 3.127

The Circumference of the Ellipse: An Empirical Abbreviation

MathCad® allows us to establish that, for m = 9 and to fourteen places of decimals, the element:-

$$1 + \sum_{n=1}^{m}\left[\frac{\prod_{k=n}^{2n-1} k}{2^{2n-1} \cdot (n)!}\right]^2 \cdot \frac{\left(\frac{(\Phi - 1)(\Phi - 1)}{(\Phi + 1)(\Phi + 1)}\right)^n}{(2n - 1)^2} = 1.01398123900053$$

Equation 3.128

from which online Wolfram® Alpha® solution using the model:-

$$\Phi^{\frac{1}{u}} = 1.01398123900053$$

Equation 3.129

gives:-

$$u = 34.65844546284822$$

Equation 3.130

The theoretical status of u, if any, is not known to me.
Given these facts we may give an empirical estimator of the Circumference (Perimeter) Length of the Golden Ellipse as:-

$$C_{empirical} = \pi\left(\frac{\Phi + 1}{2}\right) \cdot \left[\Phi^{\frac{1}{u}}\right] = 4.19689459467392$$

Equation 3.131

Please refer to Figure 4.1 that is an annotated representation of a pentagon and a pentagram inscribed within a circle. The Major Circumscribed Circle Radius, R, is unity.

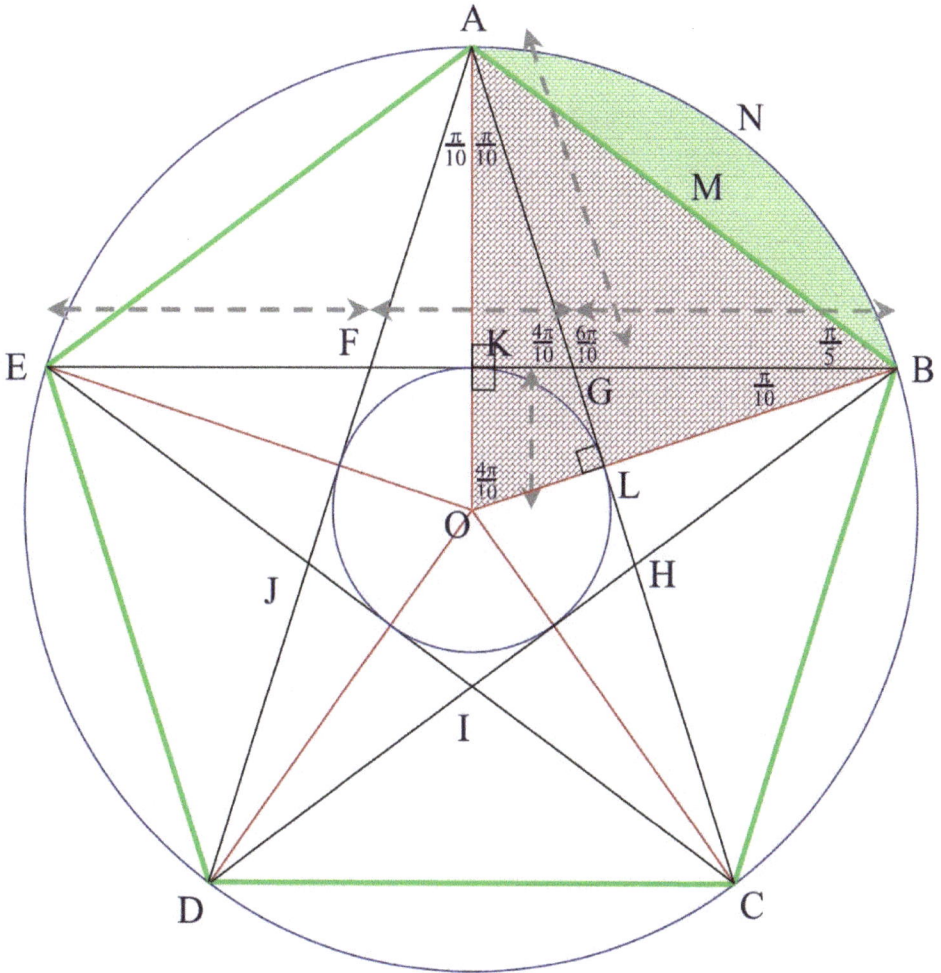

Figure 4.1
A Pentagon and a Pentagram Inscribed within a Circle

The Major Pentagon Sector Area, A_{SECTOR}

A_{SECTOR} is the Area of Sector OA(N)B and is defined as:-

$$A_{SECTOR} = \frac{\pi R^2}{5} = \cos^{-1}\left(\frac{\Phi}{5}\right)$$
Equation 4.1

The Principal Sector Triangle Area, A_{AOB}

This is the area shaded with red brick patterning in Figure 4.1, the triangle AOB defined as:-

$$A_{AOB} = \frac{1}{2}RR\sin\left(\frac{4\pi}{10}\right) = \frac{1}{2}R^2\sin\left(\frac{2\pi}{5}\right)$$
Equation 4.2

From which me may adjudge that:-

$$A_{AOB} = \frac{1}{2}RR\sin\left(\frac{\sqrt{10+2\sqrt{5}}}{4}\right) = \frac{1}{8}R^2\sqrt{2(\phi^3+7)}$$
Equation 4.3

Implied Major Segment Area

This is the segment A(M)BN of Figure 4.1 figured in a green imbrication. It is given by:-

$$A_{ImpSeg} = A_{SECTOR} - A_{AOB}$$
Equation 4.4

then substitution:-

$$A_{ImpSeg} = \frac{\pi R^2}{5} - \frac{R^2\sqrt{2(\phi^3+7)}}{8} = R^2\left(\frac{\pi}{5} - \sqrt{2(\phi^3+7)}\right)$$
Equation 4.5

Equivalation of the Whole Regular Pentagon Area[4.1]

Note that the Pentagon Apical Radius, R, is identical to the circumscribed Major Circle Radius, R. With reference to Figure 4.1, it is accordingly the case that line elements a = b = OA = OB = 1.

The Number of Pentagon Sides, n (local symbol), is 5 and The Minor Phidian Ratio, ϕ, is 0.618033988749895

Accordingly:-

$$\phi = \sqrt{a^2 + b^2 - 2ab\cos\left(\frac{\pi}{5}\right)}$$

Equation 4.6

Regular Pentagon Edge Length, t

Noting that R = 1 and n = 5, we may note by the Cosine Rule that:

$$t = \sqrt{2R^2\left(1 - \cos\left(\frac{2\pi}{5}\right)\right)} = \frac{10.R}{\sqrt{50 + 10\sqrt{5}}}$$

Equation 4.7

From which it follows that:-

$$t = R\sqrt{\frac{5 - \sqrt{5}}{2}} = \sqrt{\phi^2 + 1} = \eta$$

Equation 4.8

The Great Diagonal = AC = AD = EB

By the Cosine Rule:-

$$t = \sqrt{2R^2\left(1 - \cos\left(\frac{8\pi}{10}\right)\right)} = \sqrt{2R^2\left(1 - \cos\left(\frac{4\pi}{5}\right)\right)}$$

Equation 4.9

Then because:-

$$\cos\left(\frac{4\pi}{5}\right) = -\frac{\Phi}{2}$$

Equation 4.10

It follows that:-

$$AC = \sqrt{2R^2\left(1 - \frac{-\Phi}{2}\right)} = \sqrt{2R^2\left(1 + \frac{\Phi}{2}\right)}$$

Equation 4.11

Accordingly:-

$$\frac{AC^2}{2.R^2} = 1 + \frac{\Phi}{2}$$

Equation 4.12

And so:-

$$\Phi = 2\left(\frac{AC^2}{2.R^2} - 1\right)$$

Equation 4.13

Regular Pentagon Areas

We now discuss the Major Regular Pentagon shown in Figure 4.2 as ABCDE. Allow that the *AREA* of Pentagon ABCDE is A, and further note that:-

$$\frac{5 + \sqrt{5}}{2} = 2 + \Phi$$

Equation 4.14

Then:-

$$A = \frac{n}{2}R^2 \sin\left(\frac{2\pi}{5}\right) = \frac{5}{4}R^2\sqrt{\frac{5 + \sqrt{5}}{2}} = \frac{5}{4}R^2\sqrt{\Phi + 2}$$

Equation 4.15

The Basal Angle, AOB = θ

$$\theta = \frac{2\pi}{5}$$

Equation 4.16

and:-

$$\cos\theta = \frac{a^2 + b^2 - t^2}{2ab} = \frac{t^2 - a^2 - b^2}{-2ab}$$

Equation 4.17

or:-

$$\cos\theta = \frac{2R^2 - t^2}{2R^2} = \frac{t^2 - 2R^2 - b^2}{-2R^2}$$

Equation 4.18

The Half Basal Angle, FAG = ABK = η

$$\eta = \frac{\pi}{5}$$

Equation 4.19

and:-

$$\cos\theta = \frac{a^2 + b^2 - \left(\frac{t}{2}\right)^2}{2ab} = \frac{\left(\frac{t}{2}\right)^2 - a^2 - b^2}{-2ab}$$

where locally a = b = R. So:-

$$\cos\theta = 1 - \frac{t^2}{2R^2}$$

Equation 4.20

Pentagon Identities involving ϕ and Φ

$$\Phi = 2(1 + \phi^2)\left[1 - \cos\left(\frac{3\pi}{5}\right)\right] - 2 = 2\left\{(1 + \phi^2)\left[1 - \cos\left(\frac{3\pi}{5}\right)\right] - 1\right\}$$

Equation 4.21

Now:-

$$1 - \cos\left(\frac{3\pi}{5}\right) = \frac{\Phi + 2}{2(1 + \phi^2)}$$

Equation 4.22

from which we may write:-

$$\cos\left(\frac{3\pi}{5}\right) = 1 - \frac{\Phi + 2}{2(1 + \phi^2)} = 1 - \frac{\Phi + 2}{2[1 - (\Phi^2 - 2 * \Phi + 1)]}$$

Equation 4.23

$$\cos\left(\frac{3\pi}{5}\right) = 1 - \frac{\Phi + 2}{2(1 + \phi^2)} = 1 - \frac{\Phi + 2}{2(\Phi^2 - 2 * \Phi + 2)}$$

Equation 4.24

Knowing this we may move forward with:-

$$\frac{3\pi}{5} = \cos^{-1}\left(1 - \frac{\Phi + 2}{2(1 + \phi^2)}\right)$$
Equation 4.25

From which it is obvious that:-

$$\pi = \frac{5}{3} \cdot \cos^{-1}\left(1 - \frac{\Phi + 2}{2(1 + \phi^2)}\right)$$
Equation 4.26

Figure Analysis

Reference to Figure 4.1 shall confirm that:-

$$KB = \frac{AC}{2}$$
Equation 4.27

whilst:-

$$AB = t$$
Equation 4.28

In a similar way:-

$$\cos\left(\frac{\pi}{5}\right) = \frac{KB}{AB} = \frac{\sqrt{5} + 1}{4} = \frac{AC}{2t} = \sqrt{\frac{3 + \sqrt{5}}{8}}$$
Equation 4.29

whilst:-

$$\sqrt{\frac{3 + \sqrt{5}}{8}} = \frac{\sqrt{\frac{5 + \sqrt{5}}{8}}}{2\sqrt{\frac{5 - \sqrt{5}}{8}}}$$
Equation 4.30

A Pi-free Treatment of the Major Segment Area[4.2]

With reference to Figure 4.1 please note the following definitions:-

Symbol Element	Description	Equation	
ρ	Major Triangle Height	R.cos($2\pi/10$)	OM
R	Pentagon Vertex Radius	1	OA
h	Segment Height	R-r	MN
θ	Basal Angle	$4\pi/10$	AOB
a ($\eta = t$)	Chord Length		AB
s	Segment Arc Length		A(N)B

Table 4.1
Parameter Definitions

In these terms it is possible to declare:-

$$\rho = R.\cos\left(\frac{2\pi}{10}\right) = R.\cos\left(\frac{\pi}{5}\right) = R.\frac{1+\sqrt{5}}{4} = \frac{\Phi R}{2}$$

Equation 4.31

$$h = R - \rho = R - R\frac{1+\sqrt{5}}{4} = R\left(1 - \frac{1+\sqrt{5}}{4}\right) = R\left(\frac{3-\sqrt{5}}{4}\right)$$

Equation 4.32

and it follows that:-

$$\theta = \frac{4\pi}{10} = \frac{2\pi}{5}$$

Equation 4.33

also that:-

$$s = R\theta = R\frac{2\pi}{5}$$
Equation 4.34

and:-

$$AB = a = t = \eta = \psi = 2\sqrt{h(2R - h)} = r\sqrt{\frac{5 - \sqrt{5}}{2}} = \sqrt{\phi^2 + 1}$$
Equation 4.35

Allow that the Major Segment is Area A(N)B, the green-hatched area in Figure 4.1, and that its symbol is A_{MS}. Then:-

$$A_{MS} = R^2.\cos^{-1}\left(\frac{R - h}{R}\right) - (R - h)\sqrt{2Rh - h^2}$$
Equation 4.36

from which we may develop:-

$$A_{MS} = R^2.\cos^{-1}\left(1 - \frac{h}{R}\right) - (R - h)\sqrt{2Rh - h^2}$$
Equation 4.37

and by substitution for h followed by simplifications:-

$$A_{MS} = R^2.\cos^{-1}\left(1 - \frac{R\left(\frac{3 - \sqrt{5}}{4}\right)}{R}\right)$$

$$- \left\{R - R\left(\frac{3 - \sqrt{5}}{4}\right)\right\}\sqrt{2RR\left(\frac{3 - \sqrt{5}}{4}\right) - R\left(\frac{3 - \sqrt{5}}{4}\right)^2}$$

$$= R^2 . \cos^{-1}\left(1 - \frac{3 - \sqrt{5}}{4}\right)$$

$$- R\left(1 - \frac{3 - \sqrt{5}}{4}\right)\sqrt{2R^2\left(\frac{3 - \sqrt{5}}{4}\right) - R^2\left(\frac{3 - \sqrt{5}}{4}\right)^2}$$

$$= R^2 . \cos^{-1}\left(1 - \frac{3 - \sqrt{5}}{4}\right)$$

$$- R\left(1 - \frac{3 - \sqrt{5}}{4}\right)R\sqrt{2\left(\frac{3 - \sqrt{5}}{4}\right) - \left(\frac{3 - \sqrt{5}}{4}\right)^2}$$

$$= R^2 . \cos^{-1}\left(\frac{1 + \sqrt{5}}{4}\right) - R\left(\frac{1 + \sqrt{5}}{4}\right)R\sqrt{2\left(\frac{3 - \sqrt{5}}{4}\right) - \left(\frac{3 - \sqrt{5}}{4}\right)^2}$$

$$= R^2 . \cos^{-1}\left(\frac{1 + \sqrt{5}}{4}\right) - R\left(\frac{1 + \sqrt{5}}{4}\right)R\sqrt{\frac{5 - \sqrt{5}}{8}}$$

$$= R^2 . \cos^{-1}\left(\frac{\Phi}{2}\right) - R^2\left(\frac{\Phi}{2}\right)\sqrt{\frac{5 - \sqrt{5}}{8}}$$

$$A_{MS} = R^2\left[\cos^{-1}\left(\frac{\Phi}{2}\right) - \left(\frac{\Phi}{2}\right)\sqrt{\frac{5 - \sqrt{5}}{8}}\right]$$

Equation 4.38

Equivalation[4.3]

Allow that A_{ImpSeg} is the Implied Segment Area. We may reasonably think that this parameter is equivalent to the Major Segment Area A_{MS} defined above:-

$$A_{ImpSeg} = A_{MS}$$

Equation 4.39

Thus:-

$$R^2 \left[\frac{\pi}{5} - \frac{\sqrt{2(\phi^3 + 7)}}{8} \right] = R^2 \left[\cos^{-1}\left(\frac{\Phi}{2}\right) - \left(\frac{\Phi}{2}\right)\sqrt{\frac{5 - \sqrt{5}}{8}} \right]$$

Equation 4.40

and by cancelling R:-

$$\frac{\pi}{5} - \frac{\sqrt{2(\phi^3 + 7)}}{8} = \cos^{-1}\left(\frac{\Phi}{2}\right) - \left(\frac{\Phi}{2}\right)\sqrt{\frac{5 - \sqrt{5}}{8}}$$

Equation 4.41

so that:-

$$\pi_{implied} = 5 \left[\cos^{-1}\left(\frac{\Phi}{2}\right) - \left(\frac{\Phi}{2}\right)\sqrt{\frac{5 - \sqrt{5}}{8}} + \frac{\sqrt{2(\phi^3 + 7)}}{8} \right]$$

Equation 4.42

Since:-

$$\frac{\Phi}{2} = \frac{\phi + 1}{2}$$

Equation 4.43

We may write:-

$$\pi_{implied} = 5 \left[\cos^{-1}\left(\frac{\phi + 1}{2}\right) - \left(\frac{\phi + 1}{2}\right)\sqrt{\frac{5 - \sqrt{5}}{8}} + \frac{\sqrt{2(\phi^3 + 7)}}{8} \right]$$

Equation 4.44

from which:-

$$\pi_{implied} = 5 \left[\cos^{-1}\left(\frac{\phi + 1}{2}\right) + \frac{\sqrt{\phi^3 + 7}}{4.\sqrt{2}} - \frac{1}{4}(\phi + 1)\sqrt{\frac{5 - \sqrt{5}}{2}} \right]$$

$$\pi_{implied} = 5\left[\cos^{-1}\left(\frac{\phi+1}{2}\right) + \frac{\sqrt{\phi^3+7}}{2^{\frac{5}{2}}} - \frac{1}{4}(\phi+1)\sqrt{\frac{5-\sqrt{5}}{2}}\right]$$

$$\pi_{implied} = 5\left[\cos^{-1}\left(\frac{\phi+1}{2}\right) + \frac{\sqrt{\phi^3+7}}{\sqrt{32}} - \frac{1}{4}(\phi+1)\sqrt{\frac{5-\sqrt{5}}{2}}\right]$$

$$\pi_{implied} = 5\left[\cos^{-1}\left(\frac{\phi+1}{2}\right) + \sqrt{\frac{\phi^3+7}{32}} - \frac{\phi+1}{\sqrt{32}}\sqrt{5-\sqrt{5}}\right]$$

$$\pi_{implied} = 5\left\{\cos^{-1}\left(\frac{\phi+1}{2}\right) + \frac{1}{\sqrt{32}}\left[\sqrt{\phi^3+7} - (\phi+1)\sqrt{5-\sqrt{5}}\right]\right\}$$

Equation 4.45

Since:-

$$\sqrt{5} = 2\phi + 1$$
Equation 4.46

and:-

$$(2\phi+1)^2 = 5$$
Equation 4.47

We may continue in these terms:-

$$\pi_{implied} = 5\left\{\cos^{-1}\left(\frac{\phi+1}{2}\right) \right.$$
$$\left. + \frac{1}{\sqrt{32}}\left[\sqrt{\phi^3+7} - (\phi+1)\sqrt{(2\phi+1)^2 - (2\phi+1)}\right]\right\}$$

$$\pi_{implied} = 5\left\{\cos^{-1}\left(\frac{\phi+1}{2}\right) + \frac{1}{\sqrt{32}}\left[\sqrt{\phi^3+7} - (\phi+1)\sqrt{4\phi^2+2\phi}\right]\right\}$$

$$\pi_{implied} = 5\left\{\cos^{-1}\left(\frac{\phi+1}{2}\right) + \frac{1}{\sqrt{32}}\left[\sqrt{\phi^3+7} - \sqrt{2}(\phi+1)\sqrt{2\phi^2+\phi}\right]\right\}$$

$$\pi_{implied} = 5\left\{\cos^{-1}\left(\frac{\phi+1}{2}\right) + \frac{1}{\sqrt{32}}\left[\sqrt{\phi^3+7} - \sqrt{2}(\phi+1)\sqrt{\phi^2+1}\right]\right\}$$

Equation 4.48

At this stage the MathCad® users amongst us may note that:-

$$PSD\left(\pi, \pi_{implied}\right) = 0.000000000000014$$

and almost as interestingly:-

$$\sqrt{\phi^3 + 7} = 2.68999404785583 \approx 2.69$$

Angular Studies

Re-arrangement of Equation 4.48 enables us to write:-

$$\pi_{implied} = 5\left\{\cos^{-1}\left(\frac{\phi + 1}{2}\right) + \frac{\sqrt{\phi^3 + 7}}{\sqrt{32}} - \frac{\sqrt{2}(\phi + 1)\sqrt{\phi^2 + 1}}{\sqrt{32}}\right\}$$

Equation 4.49

Noting that the Angle α_1 is defined in these terms:-

$$\alpha_1 = 5.\cos^{-1}\left(\frac{\phi + 1}{2}\right) = 5.\cos^{-1}\left(\frac{\Phi}{2}\right) = \pi$$

Equation 4.50

The logic therefore dictates that:

$$\frac{1}{\sqrt{32}}\left[\sqrt{\phi^3 + 7} - \sqrt{2}(\phi + 1)\sqrt{\phi^2 + 1}\right] = 0$$

Equation 4.51

and:-

$$\sqrt{\phi^3 + 7} = \sqrt{2}(\phi + 1)\sqrt{\phi^2 + 1}$$

Equation 4.52

Therefore it must be that these angular factors are equivalent:-

$$\alpha_2 = \frac{\sqrt{\phi^3 + 7}}{\sqrt{32}} = \frac{1}{4}\sqrt{\frac{\phi^3 + 7}{2}} = \frac{\sqrt{\phi^3 + 7}}{4\sqrt{2}}$$

Equation 4.53

$$\alpha_3 = -\frac{\sqrt{2}(\phi + 1)\sqrt{\phi^2 + 1}}{\sqrt{32}} = \frac{(\phi + 1)\sqrt{\phi^2 + 1}}{4} = \frac{\Phi\sqrt{\phi^2 + 1}}{4}$$

Equation 4.54

and that::-

$$\pi = \alpha_1 + \alpha_2 + \alpha_3$$

Equation 4.55

Therefore, π_{implied} may be re-phrased as:-

$$\pi_{implied} = 5\left\{\cos^{-1}\left(\frac{\phi + 1}{2}\right) + \frac{1}{4}\sqrt{\frac{\phi^3 + 7}{2}} + \frac{\Phi\sqrt{\phi^2 + 1}}{4}\right\}$$

$$= 5\left\{\cos^{-1}\left(\frac{\phi + 1}{2}\right) + \frac{1}{4}\left[\sqrt{\frac{\phi^3 + 7}{2}} + \Phi\sqrt{\phi^2 + 1}\right]\right\}$$

Equation 4.56

so:-

$$\pi_{implied} = 5\left\{\cos^{-1}\left(\frac{\Phi}{2}\right) + \frac{1}{4}\left[\sqrt{\frac{\phi^3 + 7}{2}} + \Phi\sqrt{\phi^2 + 1}\right]\right\}$$

Equation 4.57

Again, if:-

$$\frac{1}{4}\left[\sqrt{\frac{\phi^3 + 7}{2}} + \Phi\sqrt{\phi^2 + 1}\right] = 0$$

Equation 4.58

then the logic dictates that:-

$$\sqrt{\frac{\phi^3 + 7}{2}} = \Phi\sqrt{\phi^2 + 1}$$

Equation 4.59

whilst:-

$$2 + \Phi = \Phi^2(\phi^2 + 1) = \frac{\phi^3 + 7}{2}$$

Equation 4.60

and:-

$$\phi^3 = 2\Phi^2(\phi^2 + 1) - 7$$
Equation 4.61

Further Powers of Φ and $z^{4.4}$

Because Φ is $\phi+1$ we may further note that:-

$$\phi^2 = (\Phi - 1)^2 = \Phi^2 - 2\Phi + 1$$
Equation 4.62

and that:-

$$\phi^3 = (\Phi - 1)^3 = \Phi^3 - 3\Phi^2 + 3\Phi - 1$$
Equation 4.63

Given the fact of Equation 4.61 we may expand to yield:-

$$\phi^3 = 2\Phi^2(\phi^2 + 1) - 7 = (2\phi^4 + 4\phi^3 + 4\phi^2 + 4\phi + 2) - 7$$
Equation 4.64

$$\phi^3 = 2\phi^4 + 4\phi^3 + 4\phi^2 + 4\phi - 5$$
Equation 4.65

which implies that:-

$$2\Phi^4 - 4\Phi^3 + 4\Phi^2 - 7 = \Phi^3 - 3\Phi^2 + 3\Phi - 1$$
Equation 4.66

or:-

$$2\Phi^4 - 5\Phi^3 + 7\Phi^2 - 3\Phi - 6 = 0$$
Equation 4.67

The Square and Square Root of the Major Ratio Φ and its Half z

$$\Phi^2 = 2 + \phi$$
Equation 4.68

$$1 - \left(\frac{\Phi}{2}\right)^2 = 1 - \left[\frac{2+\phi}{4}\right] = \frac{1}{2} - \frac{\phi}{4} = \frac{1}{4}(2 - \phi)$$
Equation 4.69

whilst:-

$$\sqrt{1 - \left(\frac{\Phi}{2}\right)^2} = \sqrt{\frac{1}{4}(2 - \phi)} = \frac{1}{2}\sqrt{2 - \phi} = \frac{\sqrt{2 - \phi}}{2}$$
Equation 4.70

The Inverse Cosine and Inverse Tangent of z

$$\cos^{-1} z = \tan^{-1}\left(\frac{\sqrt{1 - \left(\frac{\Phi}{2}\right)^2}}{\frac{\Phi}{2}}\right) = \tan^{-1}\left(\frac{2}{\Phi}\sqrt{1 - \left(\frac{\Phi}{2}\right)^2}\right)$$

Equation 4.71

so to continue:-

$$\cos^{-1} z = \tan^{-1}\left(\frac{2}{\Phi}\frac{\sqrt{2 - \phi}}{2}\right) = \tan^{-1}\left(\frac{\sqrt{2 - \phi}}{1 + \phi}\right)$$

Equation 4.72

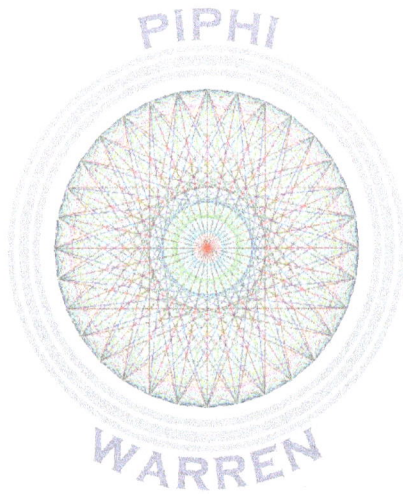

CHAPTER FIVE
THE BEHAVIOR OF FIBONACCI SERIES

You may think me a grumpy old pedant, but I do not like the epithet "Golden" applied to concepts. The word has an "almost metaphysical resonance" and fascinates with all the baleful fascination of the Incorruptible Corruptor. No, I much prefer anodyne appellations whether named for people, as in "Newton's Rings", from locations as in "India rubber", or for attributes as in "Chinstrap Penguin".

Chemists of course sneer at such "trivial" names and much prefer things like "[2(2′)E]-[2,2′-Biindolylidene]-3,3′(1H,1′H)-dione" whilst normal people prefer "indigo"; or the even more obscure "1,2-Dihydroxyanthracene-9,10-dione" which the rest of us think is madder.

It is in such a spirit that I propose to follow others in calling the "Golden Spiral" the Fibonacci Spiral just as I prefer to call things that involve Phi "Phidian" or occasionally "Phidean". For certain, neither Phidias nor Fibonacci were authors of this concept, giants though they were, because the origins of both Pi and Phi are lost in time.

Why Fibonacci?

Well, we do not know who Fibonacci really was. Like many ancient authorities Fibonacci may be a composite figure, himself a conceptual shorthand for a congeries of topics that seem to be somehow related, like "Homer", "Euclid" or "Shakespeare". The literal meaning of the Medieval Italian word is "Son of Bonacci". The word "Fibonacci" was gifted to us as late as 1838AD. Like all Italian intellectuals of the Middle Ages seem to be, the Son of Bonacci was called "Lionardo" by his near-contemporaries, though whether by nomination or description is equally unclear. Later, Fibonacci was identified with Pisan accountant Leonardo Bigollo though this figure was probably what we call in our ugly modern jargon a facilitator.

In 1202AD the *Liber Abaci* or "Book of the Abacus" appeared and its nominal author was Fibonacci. It introduced Arabic numerals to the European, Christian world. It also introduced a number of other Oriental concepts, particularly some of a mathematical character.

Fibonacci gives us a fanciful description of the breeding of rabbits. Starting with one rabbit who seems to my dull apprehension a little bit parthenogenic, Fibonacci postulates a first generation of one rabbit, a second generation of two rabbits, a third of three, a fourth of five, and so on *ad infinitum*, (though the *Liber* goes up to two hundred odd).

The Fibonacci rule of increase is:-

$$F_n = F_{n-1} + F_{n-2}$$
Equation 5.1

where F is the nth. Fibonacci Number, an integer. Fibonacci Numbers 1 and 2 are both initialised to unity.

The first nine Fibonacci Numbers are accordingly:-

1, 1, 2, 3, 5, 8, 13, 21, 34

Something else is equally true but less obvious. The quotient of successive Fibonacci Numbers, F(n)/F(n-1) approaches but never achieves the Major Ratio Of Phidias, Φ. It follows that the complimentary quotient, F(n)/F(n+1) approaches ϕ:-

$$\lim_{n \to \infty} \frac{F_{n-1}}{F_n} = \phi$$
Equation 5.2

This growth of the Fibonacci Number and convergence of the quotient is illustrated in Table 5.1

An exponential empirical equation fits the $\{n, F_n\}$ relationship and I give it below:-

$$F_n = a_{fit}e^{b_{fit}.n} = 0.459040426571e^{0.480016937188n}$$
Equation 5.3

The R^2 Correlation co-efficient for this is 0.999998601823.

We will call the Co-efficient of the RHS of Equation 5.3 a_{fit} = 0.459040426571 and the Gradient b_{fit} = 0.480016937188.

Binet's Formula

Named for and described by Jacques Philippe Marie Binet in the mid nineteenth-century, this closed-form formula was already known to Daniel Bernoulli (circa 1750), and Abraham De Moivre (circa 1710).

The Binet Formula constructs individual F_n and by implication the integer Fibonacci Series without recourse to iteration. It is given by:-

$$F_j = \frac{\Phi^j - \psi^j}{\Phi - \psi} = \frac{\Phi^j - \psi^j}{\sqrt{5}} = \frac{\Phi^j - (-\Phi)^{-j}}{\sqrt{5}} = \frac{\Phi^j - (-\Phi)^{-j}}{2\Phi - 1}$$

Equation 5.4

where ψ is the Conjugate of the Major Ratio of Phidias Φ such that:-

$$\psi = \frac{1 - \sqrt{5}}{2} = -\phi$$

Equation 5.5

It is notable that the Powers of Φ, Φ^j, and the Powers of ψ, ψ^j, though all irrational real numbers with fractional components, both form Fibonacci Series. I.e.:-

$$\Phi_j{}^j = \Phi_{j-1}{}^j + \Phi_{j-2}{}^j$$

Equation 5.6

and:-

$$\psi_j{}^j = \psi_{j-1}{}^j + \psi_{j-2}{}^j$$

Equation 5.7

Φ	1.618033989
φ	0.618033989

Serial i	Fibonacci Number F	Quotient $Q_n = (F_{n-1}/F_n)$	PSD(ϕ,Q_n)
1	1		
2	1	1	-61.80339887
3	2	0.5	19.09830056
4	3	0.666666667	-7.86893258
5	5	0.6	2.91796068
6	8	0.625	-1.12712430
7	13	0.615384615	0.42867762
8	21	0.619047619	-0.16400883
9	34	0.617647059	0.06260658
10	55	0.618181818	-0.02391930
11	89	0.617977528	0.00913553
12	144	0.618055556	-0.00348958
13	233	0.618025751	0.00133288
14	377	0.618037135	-0.00050912
15	610	0.618032787	0.00019447
16	987	0.618034448	-0.00007428
17	1597	0.618033813	0.00002837
18	2584	0.618034056	-0.00001084
19	4181	0.618033963	0.00000414
20	6765	0.618033999	-0.00000158
21	10946	0.618033985	0.00000060
22	17711	0.61803399	-0.00000023
23	28657	0.618033988	0.00000009
24	46368	0.618033989	-0.00000003
25	75025	0.618033989	0.00000001
26	121393	0.618033989	0.00000000
27	196418	0.618033989	0.00000000
28	317811	0.618033989	0.00000000
29	514229	0.618033989	0.00000000
30	832040	0.618033989	0.00000000
31	1346269	0.618033989	0.00000000
32	2178309	0.618033989	0.00000000

Table 5.1
The Behavior of the Fibonacci Sequence

<u>U Sequences</u>

Allow that:-

$$c = \frac{1}{\sqrt{5}} = 0.447213595499958$$

Equation 5.8

and:-

$$d = -\frac{1}{\sqrt{5}} = -0.447213595499958$$

Equation 5.9

then:-

$$U_j = c\Phi^j + d\psi^j$$

Equation 5.10

From which it follows that:-

$$U_j = \frac{1}{\sqrt{5}}\Phi^j - \frac{1}{\sqrt{5}}\psi^j$$

$$= \frac{1}{\sqrt{5}}\left(\Phi^j - \psi^j\right)$$

$$= \frac{1}{\sqrt{5}}\left[\left(\frac{1}{2} + \frac{\sqrt{5}}{2}\right)^j - \left(\frac{1}{2} - \frac{\sqrt{5}}{2}\right)^j\right]$$

$$= \frac{1}{\sqrt{5}}\left[\sqrt{5}.F_j\right]$$

$$\therefore U_j = F_j$$

Equation 5.11

The Exponential Form of Fibonacci Numbers

Table 5.2 summarises our current findings with regard to U-Sequences and the fitting of exponential formula parameters to the case of integer Fibonacci Series growth.

Figure 5.1 is a comparative plot of the exponentially-increasing Fibonacci Number series for the first thirty-two F_j and its EXCEL® fitted exponential regression. The resulting exponential regression equation to twelve figures is:-

$$V_j = a_{fit}.e^{b_{fit}.j} = 0.459040426571e^{0.480016937188n}$$

Equation 5.12

where V_j is the Regression Estimate of Fibonacci Number, F_j.

The analytic integrities of a_{fit} and b_{fit} (if any) are currently unknown.

Under the experimental circumstances it may be plausible that:-

$$a_{refined} = \frac{1}{\sqrt{5}} = 0.447213595499958$$

Equation 5.13

and we shall provisionally adopt that assumption.

PHIDIAN CONSTANTS

Major Ratio of Phidias	Φ	1.618033988749890
Minor Ratio of Phidias	ϕ	0.618033988749895
Conjugate of Major Ratio	ψ	-0.618033988749895

U SEQUENCE

a	$1/5^{0.5}$	0.447213595499958
b	-a	-0.447213595499958

EXPONENTIAL FITMENT

a_{fit}		0.459040426571000
b_{fit}		0.480016937188000
$a_{refined}$	$1/5^{0.5}$	0.447213595499958
$b_{refined}$	bb_{37}	0.481211825059603
$\ln(5^{0.5})$		0.804718956217050

Table 5.2
The Exponential Model Outcomes for the
Fibonacci Series

$$y = 0.459040426571e^{0.480016937188x}$$
$$R^2 = 0.999998601823$$

FIBONACCI NUMBERS   **Expon. (FIBONACCI NUMBERS)**

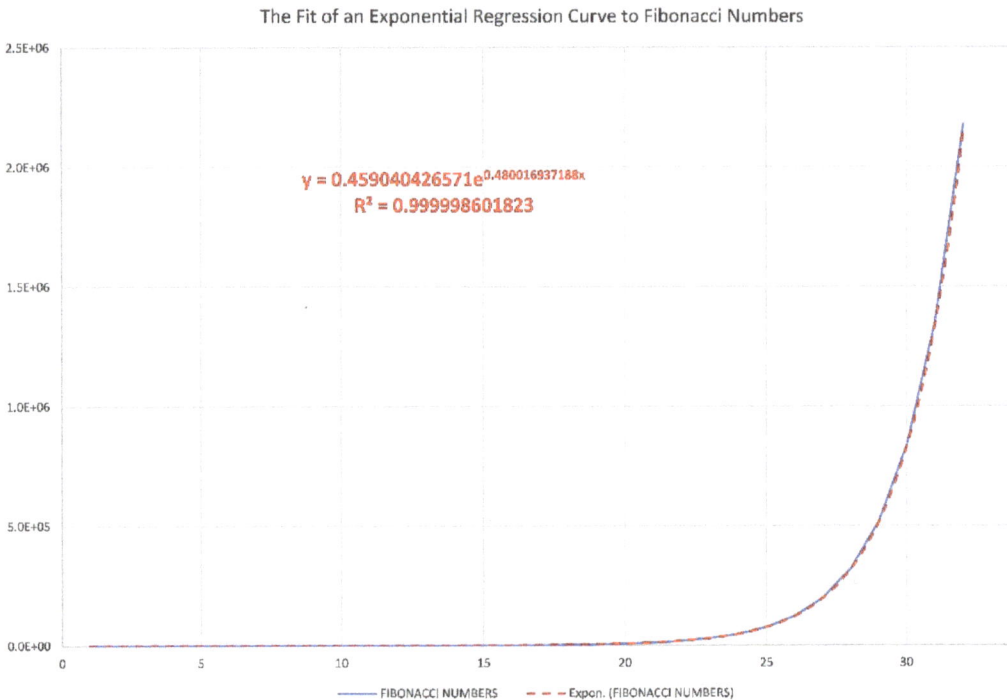

Figure 5.1
The First Thirty-Two Fibonacci Integers and their
Exponential Regression Curve

Given the $x \equiv j$ and $y \equiv F_j$ you may adjudge the fidelity of the fitted regression equation, that we shall adopt as a basis of more refined work.

Accordingly, we shall pursue a more refined relation, call it X_j, of e, the Napierian Base (Euler's Constant), to the integer Fibonacci Number, F_j.

$$F_j \approx X_j = a_{refined} \cdot e^{b_{refined} \cdot j} = \frac{1}{\sqrt{5}} \cdot e^{b_{refined} \cdot j}$$

Equation 5.14

It is further the case that:-

$$F_j \approx V_j \approx X_j = a_{refined}.\, e^{b_{refined}.j} = \frac{1}{\sqrt{5}}.\, e^{b_{refined}.j}$$

Equation 5.15

and we shall also find use for the Fractional Defects, FDV$_j$ and FDX$_j$:-

$$FDV_j = \frac{U_j - V_j}{U_j}$$

Equation 5.16a

$$FDX_j = \frac{U_j - X_j}{U_j}$$

Equation 5.16b

The approximation X$_j$ converges more rapidly to F$_j$ than does V$_j$. This can be shown by computing the Factional Residues FDV$_j$ and FDX$_j$ and comparing them. This is illustrated on Table 5.3 where it can be seen that FDV$_j$ shows monotonic decline and FDX$_j$ oscillative convergence to zero.

Serial j	FDV$_j$ = Ua$_j$ - V$_j$ over Ua$_j$	FDX$_j$ = Ua$_j$ -X$_j$ over Ua$_j$	b$_{refined}$
1	0.258143952197957	0.276393202250021	0.804718956217050
2	-0.198914875039540	-0.170820393249937	0.402359478108525
3	0.031215771678504	0.052786404500042	0.499288712258999
4	-0.043769273459079	-0.021749194749951	0.475832811221290
5	-0.012102424803490	0.008065044950046	0.482831373730230
6	-0.022287827735922	-0.003115294937453	0.480693416316148
7	-0.016689952078353	0.001184835788508	0.481381187668370
8	-0.017141317118633	-0.000453309249953	0.481155174242559
9	-0.015290636204280	0.000173040323576	0.481231053425912
10	-0.014320762695958	-0.000066111331771	0.481205214144952
11	-0.013016934044577	0.000025249974766	0.481214120540835
12	-0.011842523126531	-0.000009644968704	0.481211021316088
13	-0.010620736229773	0.000003684001120	0.481212108444827
14	-0.009419018060522	-0.000001407170378	0.481211724547504
15	-0.008211635203611	0.000000537490211	0.481211860892294
16	-0.007008402801732	-0.000000205303145	0.481211812228158
17	-0.005805573900903	0.000000078418801	0.481211829672474
18	-0.004604575627671	-0.000000029953321	0.481211823395530
19	-0.003404861149144	0.000000011441149	0.481211825661769
20	-0.002206636720372	-0.000000004370131	0.481211824841097
21	-0.001009821288750	0.000000001669240	0.481211825139091
22	0.000185556584791	-0.000000000637594	0.481211825030622
23	0.001379510155149	0.000000000243539	0.481211825070192
24	0.002572036721251	-0.000000000093025	0.481211825055727
25	0.003763139666569	0.000000000035531	0.481211825061025
26	0.004952820050458	-0.000000000013572	0.481211825059081
27	0.006141079816134	0.000000000005184	0.481211825059795
28	0.007327920566812	-0.000000000001981	0.481211825059533
29	0.008513344032618	0.000000000000756	0.481211825059630
30	0.009697351892466	-0.000000000000289	0.481211825059594
31	0.010879945842012	0.000000000000109	0.481211825059607
32	0.012061127567740	-0.000000000000043	0.481211825059602

Table 5.3
The Convergence of FDV$_j$ and FDX$_j$ Compared And Showing the Convergence of the Exponential Gradient b$_{refined}$

This greater accuracy of approximation substantiates our speculation that $a_{refined}$ is $1/5^{\frac{1}{2}}$.

To estimate $b_{refined}$ we transpose Equation 5.15 to obtain:-

$$\sqrt{5}.F_j = e^{b_{refined}.j}$$

Equation 5.17

Then taking logarithms and transposing:-

$$b_{refined} = \frac{\ln\left(\sqrt{5}\right) + \ln\left(F_j\right)}{j}$$

Equation 5.18

At $j = 40$ the value of $b_{refined}$ is 0.48121182(4841097)

The Napierian logarithm of $5^{\frac{1}{2}}$ is a constant, and it has the value 0.80471895611705

Estimates of $b_{refined}$ converge quickly, and by $n = 32$ the series has stabilised at 0.4812118250596(02). This figure is computed via EXCEL®. In this application it is unfortunately the case that MathCad® Express® becomes numerically unstable when $n > 22$.

Status at Large $j^{5.1}$

As already remarked, instabilities render numerical investigations inconvenient when j exceeds twenty-two. Fortunately, however, interesting parameters have already converged to some twelve-figure accuracy by this point.

So, defining "large" j as that at the limit $m = 22$ we may proceed as follows:-

Definitions

$$F_m = \frac{\Phi^m - \psi^m}{\sqrt{5}} \equiv F_{22} = 17711$$

Equation 5.19

$$bb_m = \frac{\ln(\sqrt{5})}{m} + \frac{\ln(F_m)}{m} \equiv b_{refined} = 0.481211825030622$$
Equation 5.20

$$u_m = \frac{\ln(\sqrt{5})}{m} = 0.036578134373502$$
Equation 5.21

$$v_m = \frac{\ln(F_m)}{m} = \frac{\ln\left(\frac{\Phi^m - \psi^m}{\sqrt{5}}\right)}{m} = 0.44463369065712$$
Equation 5.22

also:-

$$\Phi^m = \left(\frac{1}{2} + \frac{\sqrt{5}}{2}\right)^m = 39602.9999747494$$
Equation 5.23

The Taylor Series

$$\Phi^m = 1 + \sum_{k=1}^{m} \frac{1}{k!} \cdot m^k \cdot [\ln(\Phi)]^k$$
Equation 5.24

The Equation 5.24 Percentage Specific Defect in regard to LHS = Φ^m may be quoted as PSD(LHS,RHS) = 0.063476301416423

The Taylor Series in regard to the Conjugate Power ψ^m was not conveniently determinable, due to numerical error.

The Binomial Expansion

Given the notational convenience:-

$$\binom{m}{k} = \frac{m!}{k!\,(m-k)!}$$
Equation 5.25

We may offer:-

$$\Phi^m = \sum_{k=0}^{m} \binom{m}{k} \cdot \phi^k$$
Equation 5.26

The Equation 5.26 Percentage Specific Defect in regard to LHS = Φ^m may be quoted as PSD(LHS,RHS) = 0.000000000000073

whilst:-

$$\psi^m = (-\Phi)^m \sum_{k=0}^{m} \binom{m}{k} \cdot (-\Phi)^{-k}$$
Equation 5.27

The Equation 5.27 Percentage Specific Defect in regard to LHS = ψ^m may be quoted as PSD(LHS,RHS) = -0.076563051038448

The Phrase Φ^m -ψ^m

We have seen that one version of the Binet Equation is:-

$$F_j = \frac{\Phi^j - \psi^j}{\sqrt{5}}$$
Equation 5.4part

from which it immediately follows that:-

$$\Phi^j - \psi^j = \sqrt{5} \cdot F_j = (\phi + 1)^m$$
Equation 5.28

It is further demonstrable that:-

$$\Phi^m - \psi^m = \sum_{k=0}^{m} \binom{m}{k} \phi^k - (-\Phi)^m \sum_{k=0}^{m} \binom{m}{k} (-\Phi)^{-k}$$

Equation 5.29

From which substitution for Φ gives us:-

$$\Phi^m - \psi^m = \sum_{k=0}^{m} \binom{m}{k} \phi^k - (\phi - 1)^m \sum_{k=0}^{m} \binom{m}{k} (\phi - 1)^{-k}$$

Equation 5.30

or:-

$$\Phi^m - \psi^m = \sum_{k=0}^{m} \binom{m}{k} \phi^k - \sum_{k=0}^{m} \binom{m}{k} (\phi - 2)^k \sum_{k=0}^{m} \binom{m}{k} (\phi - 1)^{-k}$$

Equation 5.31

Approximation for Sufficiently Large m

As m becomes larger, parts of Equation 5.31 lose significance. Indeed, whilst avoiding numerical error, it is "sufficient" to set m to twenty-two. Under that condition the components of Equation 5.31 are:-

$$c_1 = \sum_{k=0}^{m} \binom{m}{k} \phi^k = 39602.9999747494$$

Equation 5.32a

and:-

$$c_2 = \sum_{k=0}^{m} \binom{m}{k} (\phi - 2)^k \sum_{k=0}^{m} \binom{m}{k} (\phi - 1)^{-k} = 0.000216193497181$$

Equation 5.32b

and so it becomes practicable to write:-

$$\Phi^m - \psi^m \approx \sum_{k=0}^{m} \binom{m}{k} \phi^k$$
Equation 5.33

This expression yields an precision specified in terms of the Percentage Specific Defect of:-

$$PSD\left(\Phi^m - \psi^m, \sum_{k=0}^{m} \binom{m}{k} \phi^k\right) = -0.000000063759421$$
Equation 5.34

So that the approximate value of F_m is given by:-

$$F_m \equiv F_{22} \approx \frac{1}{\sqrt{5}} \sum_{k=0}^{m} \binom{m}{k} \phi^k = 17711.0000112924$$
Equation 5.35

The actual 22nd Fibonacci Number is of course an integer and has the value 17711.
Thusly:-

$$\frac{(\phi + 1)^m}{\sqrt{5}} = 17711.0000112924$$
Equation 5.36

and:-

$$PSD(\Phi^m - \psi^m, (\phi + 1)^m) = -0.000000063759347$$
Equation 5.37

Accordingly, if an approximate Fibonacci Number is defined by the function FF(m,φ):-

$$FF(m, \phi) \approx \frac{1}{\sqrt{5}} \sum_{k=0}^{m} \binom{m}{k} \phi^k$$

Equation 5.38

then:-

$$FF(22, \phi) = 17711.000112924$$
$$FF(21, \phi) = 10945.9999817285$$

Thus:-

$$\Phi_{approx} = \frac{FF(22, \phi)}{FF(21, \phi)} = \Phi = 1.61803398874989$$

Equation 5.39

to within the limits of accuracy of my equipment. Therefore:-

$$\Phi \approx \Phi_{approx} = \frac{\frac{1}{\sqrt{5}} \sum_{k=0}^{m} \binom{m}{k} \phi^k}{\frac{1}{\sqrt{5}} \sum_{k=0}^{m-1} \binom{(m-1)}{k} \phi^k}$$

Equation 5.40

$$\Phi \approx \Phi_{approx} = \frac{\sum_{k=0}^{m} \binom{m}{k} \phi^k}{\sum_{k=0}^{m-1} \binom{(m-1)}{k} \phi^k}$$

Equation 5.41

Reciprocal Series of Sundry Qualities[5.2]

It is asserted that the series:-

$$\frac{\pi}{8} = \frac{1}{1.3} + \frac{1}{5.7} + \frac{1}{9.11} + \frac{1}{13.15} + \frac{1}{17.19} + \frac{1}{21.23} + \frac{1}{25.27} + \frac{1}{29.31} + \frac{1}{33.35}$$

Equation 5.42

is at least approximately true.

For sure this reciprocal series is mightily suggestive.

I decided to examine the quality of this equation or approximation at some length, in case it threw light upon the relations of π and ϕ.

The LHS $\pi/8$ has the value 0.392699081698724

The nine-term reciprocal series as given by Equation 5.42 has the value 0.385759975140479

Accordingly, the PSD($\pi/8$,Eqn5.42) is 1.76702897501775: Accurate enough to encourage; not accurate enough to satisfy.

So I identified the general term t as:-

$$t_j = \frac{1}{(1+4j)(3+4j)}$$

Equation 5.43

and the extended series as:-

$$T = \sum_{i=0}^{n} \frac{1}{(1+4i)(3+4i)} \approx \frac{\pi}{8} \approx \frac{1}{\Phi^2}$$

Equation 5.44

When I computed PSD($\pi/8$,T) with the number of terms taken to be n = 512 the percentage defect was 0.031024348010663 so that if the series does indeed converge to equivalence, then it must do so very slowly.

Meanwhile, PSD($\pi/8$,1/Φ^2) was 2.73317100204289, and seemed unimprovable beyond this worse than one fortieth shortfall.

Allowing (non-classically) that the first four Fibonacci Numbers F_0, F_1, F_2 and F_3 are respectively 0, 1, 2 and 3 and that:-

$$F_i = F_{i-1} + F_{i-2}$$

Equation 5.45

if n = 16 then:-

$$U = 1 + \sum_{i=1}^{n} \frac{(-1)^i}{F_i F_{i+1}}$$
Equation 5.46

and PSDS(π/8,U) is 2.73317100204289 which is very near to the result for $1/\Phi^2$. I found that sixteen was about the optimum number of terms for this series, given the capabilities of my machinery.

On the other hand, taking F_0, F_1, F_2, F_3, and F_4 to be respectively 1, 1, 2, 3 and 5 we may confidently declare:-

$$\phi = 1 + \sum_{i=1}^{n} \frac{(-1)^i}{F_i F_{i+1}}$$
Equation 5.47

since for n = 38 PSD(ϕ,Eqn5.47) is a mere zero, according to MathCad®.

The Stepping Reciprocal and the Hyperbolic Cosine

Allow that:-

$$W = \Phi^{\frac{1}{(\Phi+1)}}$$
Equation 5.48

which is very approximately cosh(ϕ).

To approximate the Hyperbolic Cosine cosh(ϕ) we may write the stepping reciprocal function defined as:-

$$W \approx V = 1 + \sum_{k=1}^{\frac{n}{2}+1} \frac{1}{F_{2k} F_{2k+1}} \approx \cosh(\phi)$$
Equation 5.49

When n = 512, and taking F_0, F_1, F_2, F_3, and F_4 to be respectively 1, 1, 2, 3 and 5, Equation 5.49 V is 1.19595578601751

PSD(W,V) is 0.48490798193741 and PSD(W,cosh(ϕ)) is 0.38636820169732. If we expand the first twelve terms of Equation 5.49 piecemeal we get:-

$$V = V_1 + V_2 + V_3$$
Equation 5.50

where:-

$$V_1 = \frac{1}{1.1} + \frac{1}{2.3} + \frac{1}{5.8} + \frac{1}{13.21} + \frac{1}{34.55} + \frac{1}{89.144}$$
Equation 5.51a

$$V_2 = \frac{1}{233.377} + \frac{1}{620.987} + \frac{1}{1597.2584}$$
Equation 5.51b

$$V_3 = \frac{1}{4181.6765} + \frac{1}{10946.17711} + \frac{1}{28657.46368}$$
Equation 5.51c

and then PSD(W,Eqn5.50) = 0.48490799263448, demonstrating that W and the series V are essentially the same after twelve iterations.

Powers and Products involving Fibonacci Successions:- Phase (1)

n = 16
Given that:-
$$\chi_j = \Phi^j$$
Equation 5.52

and that:-
$$\omega_{j+1} = F_j + \Phi.F_{J=1}$$
Equation 5.53

when:-
$$\omega_0 = 1$$
Equation 5.54

and further that:-

$$\psi_j = \frac{F_j + \Phi.F_{j+1}}{F_{j+1}} = \frac{F_j}{F_{J+1}} + \Phi$$

Equation 5.55

we are able to write:-

$$X = \sum_{j=0}^{n} \chi_j$$

Equation 5.56

$$\Omega = \sum_{j=0}^{n} \omega_j$$

Equation 5.57

and:-

$$\sum_{j=0}^{n} \chi_j = \sum_{j=0}^{n} \omega_j$$

Equation 5.58

so:-

$$X = \Omega$$

Equation 5.59

and:-

$$\Psi = \sum_{j=0}^{n-1} \psi_j \approx n.\psi_{n-1}$$

Equation 5.60

Given that n = 16, PSD(Ψ,n.ψ_{n-1}) = -0.898117437712536

Powers and Products involving Fibonacci Successions:- Phase (2)

Given that:-

$$\rho_j = \psi_j - \Phi = \frac{F_j}{F_{J+1}} \approx \phi \approx \rho_{n-1}$$

Equation 5.61

we are able to approximate the Sum of Minor Fibonacci Ratios, Ψ, in these terms:-

$$\Psi = \sum_{j=0}^{n-1} \rho_j \approx \left(n - \frac{1}{2}\right) \cdot \rho_{n-1} \approx \left(n - \frac{1}{2}\right) \cdot \phi$$

Equation 5.62

In regard to the approximations of Equation 5.62:-

$$PSD\left(\Psi, \left(n - \frac{1}{2}\right) \cdot \rho_{n-1}\right) = -0.098671563780709$$

Equation 5.63a

$$PSD\left(\Psi, \left(n - \frac{1}{2}\right) \cdot \phi\right) = -0.098597211167874$$

Equation 5.63b

The Residues of Powers Approximation

Except for F_0, F_1 and F_2, tabulation readily shows that:-

$$\sqrt{5} \cdot F_j - \Phi^j = \frac{\sqrt{5} \cdot F_j - N_j}{2}$$

Equation 5.64

which may be transposed for N_j as:-

$$N_j = 2 \cdot \Phi^j - \sqrt{5} \cdot F_j$$

Equation 5.65

N_j is the Nearest Integer to Φ^j, and is valid for the case of $j>1$
The Difference $\Delta\Phi_j$ is computed using:-

$$\Delta\Phi_j = \Phi^j - N_j$$
Equation 5.66

and $\Delta\Phi_j$ constitutes a rapidly convergent alternating series. The Nearest Integer to $\mathbb{R}$ is a standard programming function often denoted by NINT(x) or trunc(x+0.5), round(x) or some similar formulation. You can program it yourself using INT(x)+0.5. It should always be applied with due caution. In the case of this exercise I found it useful to set the first sixteen of the nearest integers manually, because the provided utility functions were anomalous.

$|\Delta\Phi_j|$ or ABS($\Delta\Phi_j$) is identical to the Equation 5.66 series except that where negative the values of $\Delta\Phi_j$ are rendered positive.

Table 5.4 presents Powers of the Major Phidian Ratio, Φ^j, and their Defects tabulated with respect to their own Nearest Integer, N_j.

Φ	1.618033988749890
$1/5^{0.5}$	0.447213595
Exp Gradient (EG)	-0.481211825

j	N_j	Φ^j	$(N_j-\Phi^j)$	$|N_j-\Phi^j|$	$(\Phi^j-N_j)/\Phi^j$	$(-1)^j \cdot \exp(EG \cdot j)$	Δ
0	0		-1	1	1.000000000000000		
1	2	1.618033989	0.38196601	0.381966011	-0.2360679774999790		
2	3	2.618033989	0.38196601	0.381966011	-0.1458980337503315	0.381966011	-6.57246E-12
3	4	4.236067977	-0.23606798	0.236067977	0.0557280900000841	-0.236067978	6.09288E-12
4	7	6.854101966	0.14589803	0.145898034	-0.0212862365252208	0.145898034	-5.02107E-12
5	11	11.09016994	-0.09016994	0.090169944	0.0081306187557583	-0.090169944	3.87858E-12
6	18	17.94427191	0.05572809	0.05572809	-0.0031056200151142	0.05572809	-2.87715E-12
7	29	29.03444185	-0.03444185	0.034441854	0.0011862412289642	-0.034441854	2.07175E-12
8	47	46.97871376	0.02128624	0.021286236	-0.0004531038537850	0.021286236	-1.46444E-12
9	76	76.01315562	-0.01315562	0.013155617	0.0001730702717712	-0.013155617	1.02036E-12
10	123	122.9918694	0.00813062	0.008130619	-0.0000661069961352	0.008130619	-6.9717E-13
11	199	199.005025	-0.005025	0.005024999	0.0000252506512343	-0.005024999	4.65391E-13
12	322	321.9968944	0.00310562	0.00310562	-0.0000096448756679	0.00310562	-3.00028E-13
13	521	521.0019194	-0.00191938	0.001919379	0.0000036840146920	-0.001919379	2.5353E-13
14	843	842.9988138	0.00118624	0.001186241	-0.0000014071683970	0.001186241	-1.9711E-13
15	1364	1364.000733	-0.00073314	0.000733137	0.0000005374905000	-0.000733137	7.92401E-14
16	2207	2206.999547	0.0004531	0.000453104	-0.0000002053031020	0.000453104	9.53995E-14

Table 5.4
Powers of the Major Phidian Ratio, F_j, and their Defects
Tabulated with respect to their own
Nearest Integer, N_j

Figure 5.2 is a plot of the difference $|N_j\text{-}\Phi^j|$ on the y-axis against j on the x-axis. Using EXCEL® I fitted an exponential regression to attain the following equation:-

$$N_j - \Phi^j = (-1)^j . e^{-0.481211825051j}$$

Equation 5.67

This equation is valid for j>1.

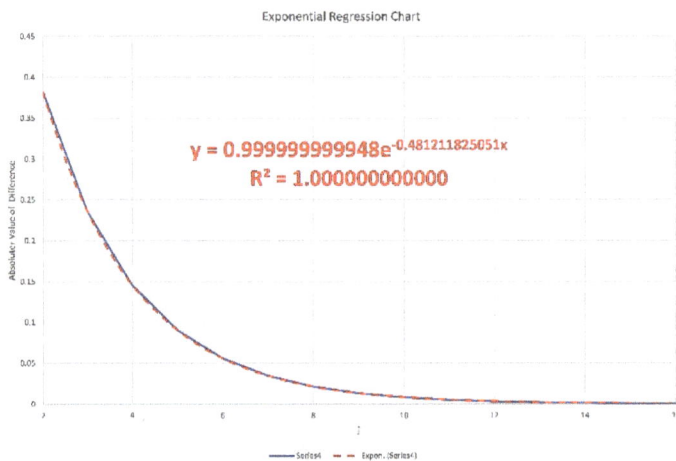

Figure 5.2 is a plot showing exponential regression chart with $y = 0.999999999948e^{-0.481211825051x}$ and $R^2 = 1.000000000000$

Figure 5.2
The Absolute Difference of $N_j - \Phi^j$
with Regard to j

Note that N_j is *not* the Fibonacci Series F_j until F_4.

CHAPTER SIX
SIMPLE SERIES SUMMATION

The concept of calculus, and of summable series such as the Taylor Series, is underpinned by the concept of summing individual areas of a function lying between ranges of the independent variable (call it x or indeed z for sake of argument).

These limited independent sub-ranges can be labeled using consecutive integers.

For example, let:-

$$\Phi = \frac{1 + \sqrt{2}}{2}$$

Equation 6.1

and:-

$$z = \frac{\Phi}{2} = \frac{1 + \sqrt{5}}{4}$$

Equation 6.2

Then, using the word "kernel" loosely, we could call the kernel of our summation for the inverse sine of z:-

$$\frac{1}{\sqrt{1 - \left(z.\frac{k}{n}\right)^2}}$$

where k is a counter going from 1 to n, i.e. k = 1...n
We can specify our sum to approximate Sin⁻¹(z) as:-

$$\sin^{-1}(z) \approx \Delta \sum_{k=1}^{n} \frac{1}{\sqrt{1 - \left(z.\frac{k}{n}\right)^2}}$$

Equation 6.3

Δ is the Interval Width, and is calculated using:-

$$\Delta = \frac{I_u - I_l}{n}$$

Equation 6.4

Where I_u is the Upper Bound of the whole range of x being summed; and I_l is the Lower Bound of the range, in our case zero.

Equation 6.3 gives a very crude estimate of the Definite Integral that defines $\text{Sin}^{-1}(z)$ so that in formal terms we may specify:-

$$\sin^{-1}(z) = \int_0^z \frac{1}{\sqrt{1-z^2}} \approx \Delta \sum_{k=1}^{n} \frac{1}{\sqrt{1-\left(z.\frac{k}{n}\right)^2}}$$

Equation 6.5

The arithmetical situation is illustrated for the case of $n = 10$ iterations by:-

Φ	1.618033989
Φ/2	0.809016994
n	10
LB=I_l	0
UB=I_u	0.809016994
Δ	0.080901699
$Sin^{-1}(Φ/2)$	0.942477796
Δ∗Σ	0.972994765
Restored Δ∗(n+Σ)	0.972994765
PSD($Sin^{-1}(Φ/2)$,Δ∗Σ)	-3.237951006

Serial	Kernel	Abstracted Kernel
1	1.003289	0.00328869
2	1.013353	0.01335294
3	1.030821	0.03082144
4	1.056868	0.05686834
5	1.093453	0.09345277
6	1.14379	0.14378978
7	1.213311	0.21331096
8	1.311805	0.31180451
9	1.458886	0.45888566
10	1.701302	0.70130162

Table 6.1

whilst Figure 6.1 represents the summation of Equation 6.3 as ten columns whose area should be summed to estimate $Sin^{-1}(z)$.

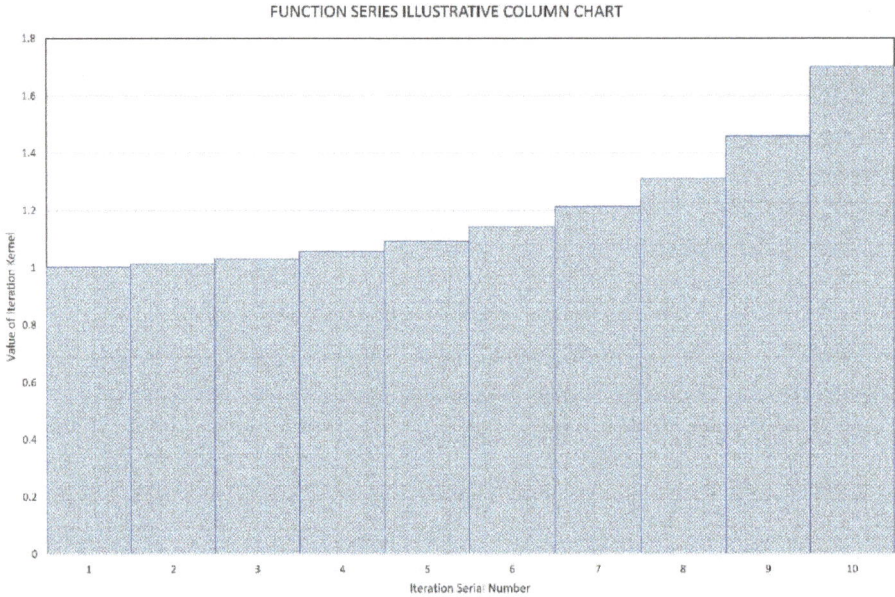

Figure 6.1

You can see straight away that there is a big block of color beneath the y = 1 level representing the area Δ×n which it would be more efficient to remove from the process and store elsewhere pending re-addition to get the final estimate. Such a removal we may call abstraction, and its re-addition we shall call restoration. This replaces n additions with one multiplication, potentially speeding and cheapening the computations, and slightly augmenting the accuracy rather than otherwise. But the actual details and advantages (if any) will depend upon the specifics of your computational methods.

So:-

$$\sin^{-1}(z) \approx \Delta \sum_{k=1}^{n} \frac{1}{\sqrt{1 - \left(z.\frac{k}{n}\right)^2}} \approx \Delta \left[n + \sum_{k=1}^{n} \left(\frac{1}{\sqrt{1 - \left(z.\frac{k}{n}\right)^2}} - 1 \right) \right]$$

Equation 6.6

Amongst the many other important features of economical and precise summation is the fact that the more and finer intervals you can add up; the more accurate your estimate. So if for example n is 64 rather than 10 we can define more and also more slender area columns in the range. Table 6.2 is the portion of the relevant worksheet that reports the inputs and results for 64 intervals applied to this problem:-

Φ	1.618033989
$\Phi/2$	0.809016994
n	64
LB=l_l	0
UB=l_u	0.809016994
Δ	0.012640891
$Sin^{-1}(\Phi/2)$	0.942477796
$\Delta*\Sigma$	0.946963368
Restored $\Delta*(n+\Sigma)$	0.946963368
PSD($Sin^{-1}(\Phi/2),\Delta*\Sigma$)	-0.475933974

Table 6.2

and the resulting areal conformation is shown by:-

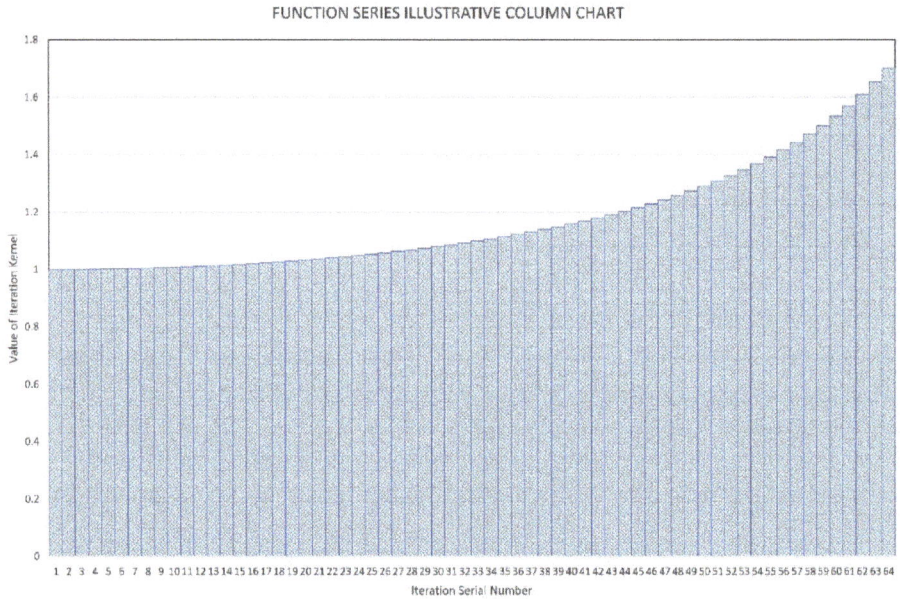

Figure 6.2

Notice how I cunningly switched n, the number of intervals, to an even power of two.

That is to facilitate the mechanical row-wise addition of a square matrix of results in order to minimise errors of addition: It is an ancient programmers' trick:-

$$64 = (2^3)^2 = 2^6$$

Notwithstanding these interesting tricks we shall move forward in a traditional way.

Since the Lower Bound, I_l, is zero, allow that I_u the Upper Bound is:-

$$u \equiv \frac{\Phi}{2}$$

Equation 6.7

and that:-

$$\Delta \equiv \frac{u}{n}$$

Equation 6.8

It is then possible by substitution to declare:-

$$\sin^{-1}(z) \approx \sum_{k=1}^{n} \frac{1}{\sqrt{1 - \left(\frac{\Phi}{2} \cdot \frac{k}{n}\right)^2}} \Delta \approx \sum_{k=1}^{n} \left(\frac{1}{\sqrt{1 - \left(u \cdot \frac{k}{n}\right)^2}} \cdot \frac{u}{n} \right)$$

Equation 6.9

and thus:-

$$\sin^{-1}(z) \approx \sum_{k=1}^{n} \left(\frac{1}{\sqrt{1 - \left(u \cdot \frac{k}{n}\right)^2}} \cdot \frac{u}{n} \right) \approx \sum_{k=1}^{n} \left(\frac{1}{\sqrt{1 - \left(u \cdot \frac{k}{n}\right)^2}} \cdot \frac{u}{n} \right)$$

$$\approx \sum_{k=1}^{n} \left(\frac{u}{n \times \sqrt{1 - \left(u \cdot \frac{k}{n}\right)^2}} \right) \approx \frac{u}{n} \sum_{k=1}^{n} \left(\frac{1}{\sqrt{1 - \left(u \cdot \frac{k}{n}\right)^2}} \right)$$

Equation 6.10

Summation for the Inverse Cosine

Recalling the identity:-

$$\cos^{-1}(z) = \frac{\pi}{2} - \sin^{-1}(z)$$

Equation 6.11

we may move forward with:-

$$X = \cos^{-1}(u)$$
Equation 6.12

and:-

$$Y = \frac{\pi}{2} - \frac{u}{n} \sum_{k=1}^{n} \left(\frac{1}{\sqrt{1 - \left(u.\frac{k}{n}\right)^2}} \right)$$

Equation 6.13

When n = 2^{12} = 4096, MathCad® Express® on my current Dell® XPS® computes:-

$$PSD(X, Y) = 0.011024880186298$$

Equation 6.14

So, crude though our method is it has already established $\text{Cos}^{-1}(z)$ to better than one part in 9000.

Towards specifying a relationship between π and Φ we can now set down:-

$$C2 = \sum_{k=1}^{n} \frac{\prod_{i=1}^{k}(2i-1)}{\prod_{j=1}^{k}(2j)} . \frac{u^{(2k+1)}}{2k+1}$$

Equation 6.15

and:-

$$\pi_{implied} = \frac{5}{3}(\Phi + 2. C2)$$

Equation 6.16

Hence:-

$$\pi_{implied} = 5Y$$

$$\pi_{implied} = 5\left[\frac{\pi}{2} - \frac{u}{n} \sum_{k=1}^{n} \left(\frac{1}{\sqrt{1 - \left(u.\frac{k}{n}\right)^2}} \right) \right]$$

$$\pi_{implied} = \frac{5}{2}\left[\pi - \frac{\Phi}{n}\sum_{k=1}^{n}\left(\frac{1}{\sqrt{1-\left(\frac{\Phi}{2}\cdot\frac{k}{n}\right)^2}}\right)\right]$$

Equation 6.17

and because:-

$$\pi_{implied} = \frac{5}{3}\frac{\Phi}{n}\sum_{k=1}^{n}\left(\frac{1}{\sqrt{1-\left(\frac{\Phi}{2}\cdot\frac{k}{n}\right)^2}}\right) = \frac{2}{3}\left[\frac{5}{2}\frac{\Phi}{n}\sum_{k=1}^{n}\left(\frac{1}{\sqrt{1-\left(\frac{\Phi}{2}\cdot\frac{k}{n}\right)^2}}\right)\right]$$

Equation 6.18

We may observe that:-

$$\left(\frac{\Phi}{2}\right)^2 = \frac{2+\phi}{4} = \frac{1}{2}+\frac{\phi}{4}$$

Equation 6.19

Whilst the identity:-

$$2\phi^4 + 3\phi^3 = 1$$
Equation 6.20

contributes a substitute for unity.

Therefore, we may excursively relate π, Φ and ϕ in the following terms:-

$$\pi_{implied} = \frac{5}{3}\frac{\Phi}{n}\sum_{k=1}^{n}\left[\frac{1}{\sqrt{2\phi^4 + 3\phi^3 - \left(\frac{\Phi}{2}\right)^2\cdot\left(\frac{k}{n}\right)^2}}\right]$$

$$\pi_{implied} = \frac{5}{3}\frac{\Phi}{n}\sum_{k=1}^{n}\left[\frac{1}{\sqrt{2\phi^4 + 3\phi^3 - \left(\frac{\phi+1}{2}\right)^2\cdot\left(\frac{k}{n}\right)^2}}\right]$$

$$\pi_{implied} = \frac{5}{3}\frac{\Phi}{n}\sum_{k=1}^{n}\left[\frac{1}{\sqrt{2\phi^4 + 3\phi^3 - \left(\frac{(\phi^2 + 2\phi + 1)}{4}\right)\cdot\left(\frac{k}{n}\right)^2}}\right]$$

$$\pi_{implied} = \frac{5}{3}\frac{\Phi}{n}\sum_{k=1}^{n}\left[\frac{1}{\sqrt{2\phi^4 + 3\phi^3 - \left(\frac{k^2}{4n^2} + \frac{\phi^2 k^2}{4n^2} + \frac{\phi k^2}{2n^2}\right)}}\right]$$

$$\pi_{implied} = \frac{5}{3}\frac{\Phi}{n}\sum_{k=1}^{n}\left[\frac{1}{\sqrt{2\phi^4 + 3\phi^3 - \frac{k^2}{4n^2}(1 + \phi^2 + 2\phi)}}\right]$$

$$\pi_{implied} = \frac{5}{3}\frac{\Phi}{n}\sum_{k=1}^{n}\left[\frac{1}{\sqrt{2\phi^4 + 3\phi^3 - \frac{k^2}{2^2 n^2}(\phi + 1)^2}}\right]$$

$$\pi_{implied} = \frac{5}{3}\frac{\Phi}{n}\sum_{k=1}^{n}\left[\frac{1}{\sqrt{2\phi^4 + 3\phi^3 - \left(\frac{k}{2n}(\phi + 1)\right)^2}}\right]$$

Equation 6.21

The quartic equation under the root sign has no concise roots, but to simplify we may note the following equivalence:-

$$2\phi^4 + 3\phi^3 - \left(\frac{k}{2n}(\phi + 1)\right)^2 = \frac{4n^2\phi^3(2\phi + 3)}{4n^2} - \frac{k^2(\phi + 1)^2}{4n^2}$$

Equation 6.22

Substitution then yields:-

$$\pi_{implied} = \frac{5}{3}\frac{\Phi}{n}\sum_{k=1}^{n}\left[\frac{1}{\sqrt{\frac{4n^2\phi^3(2\phi+3)}{4n^2} - \frac{k^2(\phi+1)^2}{4n^2}}}\right]$$

$$\pi_{implied} = \frac{5}{3}\frac{\Phi}{n}\sum_{k=1}^{n}\left[\frac{1}{\sqrt{\frac{1}{4n^2}\cdot(4n^2\phi^3(2\phi+3) - k^2(\phi+1)^2)}}\right]$$

$$\pi_{implied} = \frac{5}{3}\frac{\Phi}{n}\sum_{k=1}^{n}\left[\frac{1}{\frac{1}{2n}\cdot\sqrt{(4n^2\phi^3(2\phi+3) - k^2(\phi+1)^2)}}\right]$$

$$\pi_{implied} = \frac{5}{3}\frac{\Phi}{n}\sum_{k=1}^{n}\left[\frac{1}{\sqrt{\left(\phi^3(2\phi+3) - \frac{k^2(\phi+1)^2}{4n^2}\right)}}\right]$$

Equation 6.23

Since:-

$$\phi^3(2\phi+3) = 1$$

Equation 6.24

Appropriate back-substitution and gathering of squares returns us to an arrangement of Equation 6.18:-

$$\pi_{implied} = \frac{5}{3}\frac{\Phi}{n}\sum_{k=1}^{n}\left(\frac{1}{\sqrt{1 - \Phi^2\left(\frac{k}{2n}\right)^2}}\right)$$

Equation 6.25

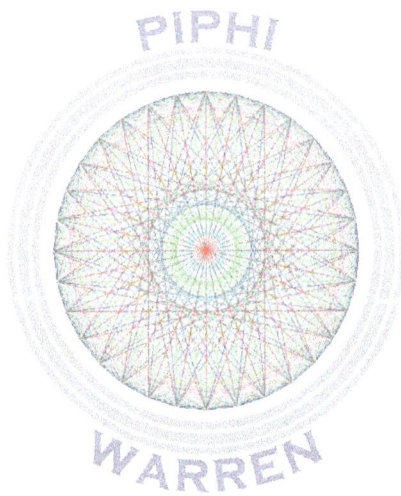

PIPHI
WARREN

CHAPTER SEVEN
INVERSE TRIGONOMETRIC FUNCTIONS

PART ONE
THE BOUNDING OF INVERSE TRIGONOMETRIC FUNCTIONS[7.1,7.2]

A distinct avenue to the approximation of inverse trigonometric functions is afforded by various derivations of the Lower Bound, LB, and the Upper Bound, UB, of solutions.

UB and LB are often of rational form and define a range in which the "true" value of the angle lies.

Numerous workers have established various values of LB and UB for $Cos^{-1}(x)$, $Sin^{-1}(x)$, and more usually $Tan^{-1}(x)$. As you might hope, there has been a tendency for these range estimates to narrow through time.

For example, for $z = \Phi/2$, we may seek:-

$$LB < \cos^{-1}(z) < UB$$
Inequality 7.1

A reasonable student might think that some average of LB and UB would necessarily yield a better estimate of the inverse function than either of the calculated bounds. But again extreme caution is needed since neither the arithmetic nor the geometric mean of the two estimates may be accurate: In fact averages may positively be misleading.

Range bounding is of course very useful for the economisation of more precise, usually iterative, methods but nevertheless great caution is needed in the assessment of the stability and precision of range bounding methods.

The Shafer-Fink Inequality

The Shafer-Fink Inequality assesses a target Inverse Sine in these terms:-

$$LB < \sin^{-1}(x) < UB$$
Inequality 7.2

with:-

$$LB = \frac{3x}{2 + \sqrt{1 - x^2}}$$
Equation 7.1

$$UB = \frac{\pi x}{2 + \sqrt{1 - x^2}}$$
Equation 7.2

In my trial of this bounding for $\text{Sin}^{-1}(z)$, I established the MathCad® Express® fiducial value of the inverse sine, AS_{fido}, as 0.942477796076938, and LB as 0.937887323136555 with UB being 0.9821553308086932

Working with these figures, the Bounds' Arithmetic Mean, μ, is 0.960020315611744 and the Bound's Geometric Mean, g, is 0.959765146810075

Accordingly the Percentage Specific Defects of the bounds, $PSD(AS_{fido},LB)$ and $PSD(AS_{fido},UB)$ are respectively 0.487064306394368 and negative 4.20970257072831. We hope for and have alternate signs (no pun intended and I trust none taken) but the wide asymmetry and the large absolute discrepancies show that this is no useful estimator of the inverse sine.

Guo Et Al: Their Bounds for $\text{Sin}^{-1}(x)$

The Guo Inequality is a sophistication of the Shafer-Fink Inequality hypothecated on the constants $\alpha = 1.752215343$ and $\beta = (\alpha\pi)/2 = 2.75237342453806$

The formal statement is:-

$$LB = \frac{(\alpha + 1)x}{\alpha + \sqrt{1 - x^2}}$$
Equation 7.3

$$UB = \frac{\beta x}{\alpha + \sqrt{1 - x^2}}$$
Equation 7.4

$$\frac{(\alpha + 1)x}{\alpha + \sqrt{1 - x^2}} < \sin^{-1}(x) < \frac{\beta x}{\alpha + \sqrt{1 - x^2}}$$
Inequality 7.3

Interestingly, for the inverse sine of z the discrepancy UB-LB is only 0.000054654110364 but the PSD(asin(x),(LB.UB)$^{\frac{1}{2}}$) is minus 0.963740757176681

D'Aurizio's Range for the Inverse Tangent[7.3]

As specified in D'Aurizio's paper:-

$$UB = \frac{\pi x\left[\left(4 + \sqrt{2}\right)\left(1 + \sqrt{1 + x^2}\right) - x\sqrt{2}\right]}{8 \times \left[1 + \sqrt{1 + x^2}\right]^2}$$
Equation 7.5

$$LB = -\frac{\pi x\left[\left(4 + \sqrt{2}\right)\left(1 + \sqrt{1 + x^2}\right) - x\sqrt{2}\right]}{8 \times \left[1 + \sqrt{1 + x^2}\right]^2}$$
Equation 7.6

where UB and LB are respectively 0.682817528091348 and negative 0.682817528091348 for the inequality:-

$$LB < \tan^{-1}(x) < UB$$
Inequality 7.4

given that x = Φ/2 =z.
If the fiducial (MathCad) value of tan^{-1}(z) is 0.680214976880921, we my assess the range of the true arc-cosine in these terms:-

$$UB = \frac{\pi x\left[\left(4 + \sqrt{2}\right)\left(1 + \sqrt{1 + x^2}\right) - x\sqrt{2}\right]}{8 \times \left[1 + \sqrt{1 + x^2}\right]^2}$$
$$- \tan^{-1}(z) = 0.002602551210427$$
Equation 7.7

$$LB = -\frac{\pi x\left[(4 + \sqrt{2})(1 + \sqrt{1 + x^2}) - x\sqrt{2}\right]}{8 \times \left[1 + \sqrt{1 + x^2}\right]^2}$$
$$+ \tan^{-1}(z) = -0.002602551210427$$

Equation 7.8

This shows that whilst the window of possibility has drastically narrowed beyond the sixty-year-old Shafer-Fink Inequality it is also the case that the true solution is highly-skewed from any mean position.

In order to "symmetricise" the range bounds we may introduce the arbitrary local co-divisor λ such that:-

$$8 \approx 2^\lambda$$
Equation 7.9

When λ = 3.00654325 the bounds become:-

$$LB = \frac{\pi x\left[(4 + \sqrt{2})(1 + \sqrt{1 + x^2}) - x\sqrt{2}\right]}{2^\lambda \times \left[1 + \sqrt{1 + x^2}\right]^2}$$
$$- \tan^{-1}(z) = -0.000487311270313$$

Equation 7.10

$$UB = -\frac{\pi x\left[(4 + \sqrt{2})(1 + \sqrt{1 + x^2}) - x\sqrt{2}\right]}{2^\lambda \times \left[1 + \sqrt{1 + x^2}\right]^2}$$
$$+ \tan^{-1}(z) = -0.000487311270313$$

Equation 7.11

Therefore the transformed D'Aurizio Bounds become:-

$$UB = \frac{\pi x\left[(4 + \sqrt{2})(1 + \sqrt{1 + x^2}) - x\sqrt{2}\right]}{2^\lambda \times \left[1 + \sqrt{1 + x^2}\right]^2} = 0.679727665610608$$

Equation 7.12

$$LB = -\frac{\pi x\left[\left(4+\sqrt{2}\right)\left(1+\sqrt{1+x^2}\right)-x\sqrt{2}\right]}{2^\lambda \times \left[1+\sqrt{1+x^2}\right]^2} = -0.679727665610608$$
Equation 7.13

The Range is now:-

$$\rho = |LB| + |UB| = 1.35945533122122$$
Equation 7.14

and it is now the case that the Means μ and η are:-

$$\mu = \frac{|LB| + |UB|}{2} = 0.679727665610608$$
Equation 7.15

$$\eta = \sqrt{|LB| \times |UB|} = 0.679727665610608$$
Equation 7.16

Clearly, it is a matter of indifference whether the Arithmetic or the Geometric Mean is applied since:-

$$\mu = \eta = \frac{\rho}{2}$$
Equation 7.17

and predictably:-

$$PSD(\tan^{-1}(z), \mu) = 0.071640773413652$$
Equation 7.18

$$PSD(\tan^{-1}(z), \eta) = 0.071640773413652$$
Equation 7.19

This demonstrates that the λ-modified D'Aurizio system brings us within a thousandth part of the solution for tan⁻¹(z) from which cos⁻¹(z) may be calculated to a similar accuracy.

<u>The Simplification of the D'Aurizio System</u>

Fragment A

$$A = 1 + \sqrt{1 + \left(\frac{\Phi}{2}\right)^2}$$

Equation 7.20

Therefore:-

$$A = 1 + \sqrt{1 + \left(\frac{\Phi^2}{4}\right)}$$

$$= 1 + \sqrt{1 + \left(\frac{2 + \phi}{4}\right)}$$

$$= 1 + \sqrt{1 + \frac{1}{2} + \frac{\phi}{4}}$$

$$= 1 + \sqrt{\frac{3}{2} + \frac{\phi}{4}}$$

$$= 1 + \sqrt{\frac{1}{2}\left(3 + \frac{\phi}{2}\right)}$$

$$= 1 + \sqrt{\frac{1}{2}}\sqrt{3 + \frac{\phi}{2}}$$

Equation 7.21

and because:-

$$\sqrt{\frac{1}{2}} = \frac{1}{\sqrt{2}}$$

Equation 7.22

we may proceed to simplify Fragment B.

Fragment B

$$B = A^2 = \left(1 + \sqrt{1 + \left(\frac{\Phi^2}{4} \right)} \right)^2$$

Equation 7.23

Therefore:-

$$B = \left[1 + \frac{1}{\sqrt{2}} \sqrt{3 + \frac{\phi}{2}} \right]^2$$

Equation 7.24

For local purposes allow that a = 1 and b is as defined below:-

$$b = \frac{1}{\sqrt{2}} \sqrt{3 + \frac{\phi}{2}}$$

Equation 7.25

then:-

$$(a + b)^2 = a^2 + 2ab + b^2$$
Equation 7.26

Substitution in the Estimate of the Inverse Cosine

So by substitution (and remembering that λ is 3.00654325):-

$$UB = \frac{\pi . \frac{1+\phi}{2} . \left[(4+\sqrt{2})\left(1+\frac{1}{\sqrt{2}}\sqrt{3+\frac{\phi}{2}}\right) - \sqrt{2}.\frac{1+\phi}{2} \right]}{2^{\lambda}\left(\frac{5}{2}+\frac{2}{\sqrt{2}}\sqrt{3+\frac{\phi}{2}}+\frac{\phi}{4}\right)}$$

Equation 7.27

$$LB = -\frac{\pi . \frac{1+\phi}{2} . \left[(4+\sqrt{2})\left(1+\frac{1}{\sqrt{2}}\sqrt{3+\frac{\phi}{2}}\right) - \sqrt{2}.\frac{1+\phi}{2} \right]}{2^{\lambda}\left(\frac{5}{2}+\frac{2}{\sqrt{2}}\sqrt{3+\frac{\phi}{2}}+\frac{\phi}{4}\right)}$$

Equation 7.28

At this juncture it is clear that UB and LB are numerically-identical except for their signs; and also that departure from the precise inverse function is controlled by lambda.

Accordingly it is clear and concise to define E, the Estimate of the Inverse Cosine of Argument Φ/2 as:-

$$E = \frac{\pi . \frac{1+\phi}{2} . \left[(4+\sqrt{2})\left(1+\frac{1}{\sqrt{2}}\sqrt{3+\frac{\phi}{2}}\right) - \sqrt{2}.\frac{1+\phi}{2} \right]}{2^{\lambda}\left(\frac{5}{2}+\frac{2}{\sqrt{2}}\sqrt{3+\frac{\phi}{2}}+\frac{\phi}{4}\right)}$$

Equation 7.29

Noting that:-

$$(4+\sqrt{2})\left(1+\frac{1}{\sqrt{2}}\sqrt{3+\frac{\phi}{2}}\right) = \frac{1}{2}.(4+\sqrt{2}).(2+\sqrt{\phi+6})$$

Equation 7.30

we may move forward with:-

$$4 + 2^{1.5} \cdot \sqrt{3 + \frac{\phi}{2}} + \sqrt{3 + \frac{\phi}{2} + \sqrt{2}} = 4 + 2\sqrt{2}\sqrt{3 + \frac{\phi}{2}} + \sqrt{3 + \frac{\phi}{2} + \sqrt{2}}$$

Equation 7.31

Because parenthetically:-

$$\log_2\left(\frac{5}{2}\right) = 1.32192809488736$$

Equation 7.32

and:-

$$2^{1.32192809488736} = 2.5$$

Equation 7.33

Further rationalisations enable:-

$$E = \frac{2^{-1}\pi(1 + \phi)\left[2^{-1}\left(4 + \sqrt{2}\right)\left(2 + \sqrt{\phi + _6}\right) - 2^{-0.5}(1 + \phi)\right]}{2^{\lambda}\left(2.5 + 2^{0.5}\sqrt{3 + \frac{\phi}{2}} + 2^{-2}\phi\right)}$$

Equation 7.34

Or alternatively:-

$$E = \frac{2^{-(\lambda+1)}\pi(1 + \phi)\left[\frac{\sqrt{\phi + 6}}{\sqrt{2}} + 2\sqrt{\phi + 6} + 4 + \frac{1}{\sqrt{2}}(1 - \phi)\right]}{\frac{5}{2} + \sqrt{\phi + 6} + \frac{\phi}{4}}$$

Equation 7.35

Now:-

$$1 - \phi = 1 - (\Phi - 1) = 2 - \Phi$$

Equation 7.36

$$\sqrt{\phi + 6} = \sqrt{\Phi + 5}$$

Equation 7.37

and:-

$$\frac{\sqrt{\phi + 6}}{\sqrt{2}} = \sqrt{\frac{\phi + 6}{2}} = \sqrt{\frac{\Phi + 5}{2}} = \sqrt{z + \frac{5}{2}}$$

Equation 7.38

Accordingly, Equation 7.35 may be re-expressed as:-

$$E = \frac{\frac{\pi\Phi}{2}\left[\sqrt{z + \frac{5}{2}} + 2\sqrt{\Phi + 5} + 4 + \frac{1}{\sqrt{2}}(2 - \Phi)\right]}{2^\lambda\left(\frac{5}{2} + \sqrt{\Phi + 5} + \frac{\Phi - 1}{4}\right)}$$

Equation 7.39

So that:-

$$\pi_{implied} \approx 5E \approx \frac{\frac{5\pi\Phi}{2}\left[\sqrt{z + \frac{5}{2}} + 2\sqrt{\Phi + 5} + 4 + \frac{1}{\sqrt{2}}(2 - \Phi)\right]}{2^\lambda\left(\frac{5}{2} + \sqrt{\Phi + 5} + \frac{\Phi - 1}{4}\right)}$$

Equation 7.40

Interestingly, E appears, in this conformation, to be a much better estimator of $Tan^{-1}(z)$ than it is of $Cos^{-1}(z)$ since:-

$$PSD\left(\pi, \pi_{implied}\right) = -8.18201793824324$$
$$PSD(atan\,(z), E) = 0.071640773413635$$

And yet the E of Equation 7.39 is 0.679727665610608 whilst $Cos^{-1}(z)$ is 0.628318530717959

An Improved Approach to E

Let us say for clarity that B is a Better Expression of the Arc-Cosine Estimate E.
Also, allow that:-

$$\cos^{-1}(z) = \cos^{-1}\left(\frac{\Phi}{2}\right)$$

$$= \tan^{-1}\left[\sqrt{\frac{4}{\Phi^2}\left(1 - \frac{\Phi^2}{4}\right)}\right] = 0.628318530717959$$

Equation 7.41

and that:-

$$\theta = \sqrt{\frac{4}{\Phi^2}\left(1 - \frac{\Phi^2}{4}\right)}$$

Equation 7.42

The new, suitable value of lambda is given by:-

$$\lambda = 3.00876575$$

Equation 7.43

and so:-

$$B$$

$$= \frac{\pi\sqrt{\frac{4}{\Phi^2}\left(1 - \frac{\Phi^2}{4}\right)}\left\{(4 + \sqrt{2})\left[1 + \sqrt{1 + \sqrt{\frac{4}{\Phi^2}\left(1 - \frac{\Phi^2}{4}\right)}^2}\right] - \sqrt{2}.\sqrt{\frac{4}{\Phi^2}\left(1 - \frac{\Phi^2}{4}\right)}\right\}}{2^\lambda\left[1 + \sqrt{1 + \sqrt{\frac{4}{\Phi^2}\left(1 - \frac{\Phi^2}{4}\right)}^2}\right]^2}$$

Equation 7.44

$$B = \frac{\pi\theta\{(4 + \sqrt{2})[1 + \sqrt{1 + \theta^2}] - \sqrt{2}.\theta\}}{2^\lambda[1 + \sqrt{1 + \theta^2}]^2}$$

Equation 7.45

$$B = \frac{\pi\theta\left\{(4+\sqrt{2})\left[1+\frac{2}{\Phi}\right] - \sqrt{2}.\theta\right\}}{2^\lambda\left[1+\frac{2}{\Phi}\right]^2} = 0.628369665094559$$

Equation 7.46

The Equation 7.46 conformation of the D'Aurizio bound yields the improved PSD's:-

$$PSD(\text{acos}(z), B) = -0.008138288797969$$
$$PSD(\pi, 5B) = -0.008138288797955$$

Therefore:-

$$\pi_{implied} = \frac{5\pi\theta\left\{(4+\sqrt{2})\left[1+\frac{2}{\Phi}\right] - \sqrt{2}.\theta\right\}}{2^\lambda\left[1+\frac{2}{\Phi}\right]^2} = 3.1418483254728$$

Equation 7.47

For which PSD(π,$\pi_{implied}$) = -0.00813828879794

THE TAYLOR SERIES FOR THE INVERSE COSINE

The Taylor Series Expansion for the inverse *sine* is specified by:-

$$\sin^{-1}(z) = \sum_{n=0}^{m} \frac{(2n)!}{2^{2n}.(n!)^2} \cdot \frac{z^{(2n+1)}}{2n+1}$$

Equation 7.48

And it is accordingly the case that:-

$$\cos^{-1}(z) = \frac{\pi}{2} - \sum_{n=0}^{m} \frac{(2n)!}{2^{2n}.(n!)^2} \cdot \frac{z^{(2n+1)}}{2n+1}$$

Equation 7.49

The summation is a slowly-convergent series that draws ever closer to the inverse sine but requires m = 48 iterations to reach an accuracy of eleven places of decimals.

An alternative specification of the series for Cos⁻¹(z) is given by:-

$$\cos^{-1}(z) = \frac{\pi}{2} - \left(z + \frac{1}{2}.\frac{z^3}{3} + \frac{1.3}{2.4}.\frac{z^5}{5} + \frac{1.3.5}{2.4.6}.\frac{z^7}{7}\cdots \right)$$

Equation 7.50

Which is accurate to a single decimal place for the four terms shown, and which generalises to:-

$$\cos^{-1}(z) = \frac{\pi}{2} - z - \sum_{k=1}^{m} \frac{\prod_{i=1}^{k}(2i-1)}{\prod_{j=1}^{k}(2j)} \cdot \frac{z^{2k+1}}{2k+1}$$

Equation 7.51

This gives around nine figures accuracy after m = 48 iterations. This figure improves to about thirteen figure accuracy after m = 98 iterations.

By isolating the sum of products we may write:-

$$C2 = \sum_{k=1}^{m} \frac{\prod_{i=1}^{k}(2i-1)}{\prod_{j=1}^{k}(2j)} \cdot \frac{z^{2k+1}}{2k+1}$$

Equation 7.52

such that:-

$$\cos^{-1}(z) = \frac{\pi}{2} - \frac{\Phi}{2} - C2$$

Equation 7.53

Knowing that:-

$$\frac{\pi}{2} - \frac{\pi}{5} = \frac{\Phi}{2} + C2$$

Equation 7.54

we may move forward with:-

$$\frac{3\pi}{10} = \frac{\Phi}{2} + C2$$
$$\pi = \frac{5}{3}\Phi + \frac{10}{3}C2$$
$$\pi = \frac{5}{3}(\Phi + 2.\,C2)$$

Equation 7.55

CHAPTER EIGHT
INTEGRALS

The Taylor Series is essentially a discretisation of the relevant generative integral so that we may declare:-

$$\sin^{-1}(z) = \int_0^z \frac{1}{\sqrt{1-z^2}} dz$$
Equation 8.1

And thusly:-

$$\cos^{-1}(z) = \frac{\pi}{2} - \int_0^z \frac{1}{\sqrt{1-z^2}} dz = \frac{\pi}{2} - \frac{3\pi}{10} = \frac{\pi}{5}$$
Equation 8.2

Accordingly it follows that:-

$$\pi = 5\left(\frac{\pi}{2} - \int_0^z \frac{1}{\sqrt{1-z^2}} dz\right) = \frac{5\pi}{2} - 5\int_0^z \frac{1}{\sqrt{1-z^2}} dz\right)$$
Equation 8.3

$$\pi = \int_{-1}^1 \frac{1}{(1-t^2)} dt$$
Equation 8.4

$$\frac{\pi}{2} = \int_0^1 \frac{1}{(1-t^2)} dt$$
Equation 8.5

and also:-

$$\frac{3}{2}\pi = 5\int_0^z \frac{1}{\sqrt{1-z^2}} dz$$
Equation 8.6

from which it follows that:-

$$\pi = \frac{10}{3} \int_0^z \frac{1}{\sqrt{1-z^2}} \, dz$$
Equation 8.7

General Definitions

For fiducial purposes we may re-visit some fundamental parameter definitions of use in our application of integrations to the determination of $\text{Cos}^{-1}(\Phi/2)$, or other desiderata.

Firstly:-

$$\Phi = \frac{1 + \sqrt{5}}{2}$$
Equation 8.8

$$z = \frac{\Phi}{2}$$
Equation 8.9

and:-

$$\sin^{-1}(z) = 0.942477796076938$$
Equation 8.10

whilst:-

$$\cos^{-1}(z) = 0.628318530717959$$
Equation 8.11

Accordingly the Fiducial Integral of z is given by:-:-

$$FidoInt(z) = \int_0^z \frac{1}{\sqrt{1-z^2}} \, dz = 0.942477796076938$$
Equation 8.12

In terms of series definitions:-

$$a = 0$$
Equation 8.13

$$b = z$$
Equation 8.14

$$ff(x) = \frac{1}{\sqrt{1 - x^2}}$$

Equation 8.15

$$f(j, h) = \frac{1}{\sqrt{1 - (j.h)^2}}$$

Equation 8.16

$$g(x) = \frac{1}{\sqrt{1 - \left(\frac{b - a}{2}\right)^2}}$$
Equation 8.17

As per our usual convention the Percentage Specific Defect PSD(x,y) is defined as below:-

$$PSD(x, y) = 100 \times \left(\frac{x - y}{x}\right)$$
Equation 8.18

Newton-Coates Formulae for Numerical Integration[8.1]

Newton-Coates (NC) formulae estimate the integrals of series of subject functions at points equally-spaced along the abscissa (x-axis). That is to say they approximate the integral of y_i, or if you prefer $f(x_i)$, for x = 0h, 1h, 2h, 3h, ... , nh where h = $(x_n-x_0)/n$. Since a finite range is defined by (x_n-x_0) it is clear that Newton-Coates integrals are definite.

Like all the other methods of numerical integration, often referred to as Numerical Quadrature in older literature, the NC formulae are applied where an analytic expression for the integral of a given equation is not available: This is the case with the Inverse Sine that is a condition of obtaining the Inverse Cosine, and hence the expression of one of the ways of relating Pi and Phi.

In my experience, gained in a wide diversity of applications, the Extended Simpson's Rule is the most robust and versatile NC method, though multi-panel extensions of Rule 18 are also often very useful, especially six- or eight-panel extensions covering maybe 49 or 65 data points (if obtainable).

Higher order NC formulae, beyond Rule 18, tend to implicate computational errors of rounding and truncation that vitiate accuracy and increase cost. In such situations, more sophisticated methods of numerical integration should be explored.

In general the Interval of Integration, h, is given by:-

$$h = \frac{b - a}{m}$$
Equation 8.19

where b is the local Range Upper Bound; a is the Lower Bound and m is the Number of Intervals which is one less than the number of Data Points.

Used with care, Newton-Coates formulae are capable of impressive accuracy, especially in their higher orders, such as A&S Rule 18, or in their concatenated forms such as the Composite Simpson's Rule.

The Extended Trapezoidal Rule A&S 25.4.2 p885

For m = 128 and j = 0 ... m:-

$$I_{ETR} = h.\left[\frac{f(0,h) + f(m,h)}{2} + \sum_{j=1}^{m-1} f(j,h)\right] = 0.942491057324704$$
Equation 8.20

PSD(Sin^{-1}(z),I$_{ETR}$) is -0.001407062089054 showing that the (in)accuracy of the method is about one part in 711.

The Modified Trapezoidal Rule A&S 25.4.4 p885

m is again 128.

This form of the Extended Trapezoidal Rule incorporates an additive corrector variable, Δ, intended to drive a few more significant figures of accuracy in the output integral estimate.

Δ is defined using:-

$$\Delta = \frac{h}{24} \cdot [-f(-1,h) + f(1,h) + f(m-1,h) - f(m+1,h)]$$

Equation 8.21

$$I_{METR} = h \cdot \left[\frac{f(0,h) + f(m,h)}{2} + \sum_{j=1}^{m-1} f(j,h) \right] + \Delta$$
$$= 0.942477785549976$$

Equation 8.22

PSD(Sin^{-1}(z),I$_{METR}$) is 0.000001116945403

Milne's Rule (Open-type NC Integration)[8.2]

m = 4.

Here we may digress to examine four Newton-Coates-related "witchcraft" formulas that were developed in pre-computer days to calculate acceptable estimates of an integral using a minimum of arithmetic, and according to the best traditions of Anglo-American "Goodenough Theory".

We start with Milne's Method, basically the predictor process of a predictor-corrector method for the numerical solution of First Order Ordinary Differential Equations (ODEs).

Milne's Method was actually developed by American William Edward Milne, sometime in the mid twentieth-century:-

$$I_{MILNE} = \frac{4}{3}h[2f(1,h) - f(2,h) + 2f(3,h)] = 0.934373367175281$$

Equation 8.23

This abridged (mis)application of part of Milne's Process gives PSD(Sin^{-1},I$_{MILNE}$) = 0.859906613757053

Durand's Rule

m = 5 and j = 0 ... m
Due to US naval engineer Willian F Durand, Durand's Rule was evolved sometime in the 1890s, possibly in connection with research into fluid discharge from an orifice.
It is convenient (and theoretically justifiable) to break the sequence of functions into two parts as shown:-

$$p_1 = \frac{2}{5}f(0,h) + \frac{11}{10}f(1,h) + f(2,h)$$
Equation 8.24a

$$p_2 = f(3,h) + \frac{11}{10}f(4,h) + \frac{2}{5}f(5,h)$$
Equation 8.24b

Durand's Rule is then given by:-

$$I_{DURAND} = h(p_1 + p_2) = 0.944746093131323$$
Equation 8.25

PSD(Sin^{-1}(z),I$_{DURAND}$) is -0.240673792404108, showing an error a little better than one part in four hundred.

Hardy's Rule

m = 6
$$p_1 = 28f(0,h) + 162f(1,h) + 220f(3,h)$$
Equation 8.26a

$$p_2 = 162f(5,h) + 28\frac{11}{10}f(6,h)$$
Equation 8.26b

Hardy's Rule is then given by:-

$$I_{HARDY} = \frac{h}{100}(p_1 + p_2) = 0.942549247918381$$
Equation 8.27

PSD(Sin^{-1}(z),I$_{HARDY}$) is -0.007581275839111, showing an error better than one part in ten thousand.

Weddle's Rule

$$m = 6$$

$$p_1 = f(0,h) + 5f(1,h) + f(2,h) + 6f(3,h)$$
Equation 8.28a

$$p_2 = f(4,h) + 5f(5,h) + f(6,h)$$
Equation 8.28b

Weddle's Rule is then given by:-

$$I_{WEDDLE} = \frac{3h}{10}(p_1 + p_2) = 0.942665545946219$$
Equation 8.29

PSD(Sin^{-1}(z),I$_{WEDDLE}$) is -0.019920879840636

Simpson's Rule A&S 25.4.5 p886

We now return to mainstream, extendible, Newton-Coates formulae. Arguably the most important NC formula is Simpson's Rule in its various guises. Because it approximates the curve of a two-dimensional function with a series of parabolas connecting points equally-spaced in the independent variable x it is extremely versatile, and used judiciously very accurate at minimal cost. It is also economical to program, and robust.

Thomas Simpson first published the rule in 1743, but there is evidence that a version was known to Bonaventura Cavalieri in 1639, and James Gregory knew the substance of the Rule by 1668.

The orthodox, single-panel form is:-

$$I_{SR} = \frac{h}{3}[f(0,h) + 4f(1,h) + f(2,h)] = 0.953981069463261$$
Equation 8.30

It is known that turners, coopers and others were using Equation 8.30 to estimate areas and volumes of product as early as the late eighteenth-century, and similar rules may be of even older vintage.

$PSD(Sin^{-1}(z),I_{SR}) = -1.22053521411385$

An alternate form is:-

$$I_{SR} = \frac{b-a}{6}\left[f(0,h) + 4f\left(\frac{b-a}{2}\right) + f(2,h)\right] = 0.953981069463261$$
Equation 8.31

PSD is identical for Equations 8.30 and 8.31

Extended Simpson's Rule A&S 25.4.6 p886

m = 128

Simpson's Rule can be applied to any continuous series that has an even number of intervals (i.e. n≡m is odd).

Like other true NC integration series its accuracy is greatly improved by establishing overlapping panels of the basic series pattern. In the case of Simpson's Rule an concise way of expressing this multi-panel structure is:-

$$I_{ESR} = \frac{h}{3}\left[f(0,h) + \sum_{k=1}^{m-1}\{3 - (-1)^k\}.f(k,h) + f(m,h)\right]$$
Equation 8.32

The term in curly brackets is an alternator that has the behavior of being 4 when k is odd and 2 when k is even. The "2" factor arises because the terminal points within the series are added as in the two-panel series:-

$$I_{ESR} = \frac{h}{3}[f(0,h) + 4f(1,h) + 2f(2,h) + 4f(3,h) + 2f(5,h) + f(6,h)]$$
Equation 8.33

An alternative specification of the Extended Simpson's Rule is accordingly:-

$$I_{ESR} = \frac{h}{3}\left[f(0,h) + 4\sum_{j=1}^{\frac{m}{2}} f(2j-1,h) + 2\sum_{j=1}^{\frac{m}{2}-1} f(2j,h) + f(m,h)\right]$$
Equation 8.34

Your choice of idiom will depend upon the particulars of your computational scheme.

The PSD(Sin^{-1}(z),I$_{ESR}$) for Equation 8.32 transpired to be negative 0.000000405023731 and the value of I$_{ESR}$ 0.942477799894197

Simpson's Three-Eighths Rule A&S 25.4.13 p886 (Four Part)[8.3]

m = 12

This format is sometimes suitable when iterations need to be minimised but accuracy maximised in a structure of simple coefficients. Therefore, it is suited to manual working.

Our four-part exercise for Parts p$_1$, ... ,p$_4$ implicates thirteen data points (inclusive of f(0,h)), and nine effective intervals.

$$p_1 = 3f(1,h) + 3f(2,h) + 2f(3,h)$$
Equation 8.35a
$$p_2 = 3f(4,h) + 3f(5,h) + 2f(6,h)$$
Equation 8.35a
$$p_3 = 3f(7,h) + 3f(8,h) + 2f(9,h)$$
Equation 8.35a
$$p_4 = 3f(10,h) + 3f(11,h) + 2f(12,h)$$
Equation 8.35a

The Three-Eighths Formula is then:-

$$I_{ESTER} = \frac{3h}{8}\left[f(0,h) + \sum_{j=1}^{4} p_j\right] = 0.942559487811949$$

Equation 8.36

$$PSD(Sin^{-1}(z), I_{ESTER}) = -0.008667762291198$$

Generalised Extended Simpson's Three-Eights Rule

m = 12

The simplest way of approaching a panel-wise generalisation of the Three-Eights Rule that is consistent with MathCad Express® syntax is to define three stepping series as below:-

i = 1, 4 … m - 2	**Stepping Series A**
j = 2, 5 … m - 1	**Stepping Series B**
k = 3, 6 … m - 3	**Stepping Series C**

The Generalised Extended Simpson's Three-Eighths Rule, I_{GESTER}, then becomes:-

$$I_{GESTER} = \frac{3h}{8}\left[f(0,h) + f(m,h)\right.$$
$$\left. + 3\sum_{i}^{m-2} f(i,h) + 3\sum_{j}^{m-1} f(j,h) + 2\sum_{k}^{m-3} f(k,h)\right]$$

Equation 8.37

The I_{GESTER} value is 0.942559487811949
$PSD(Sin^{-1}(z), I_{GESTER}) = -0.008667762291174$

Boole's Rule A&S 25.4.14

m = 4
Boole's Rule is sometimes called "Bode's" Rule in old literature due to an early misprint.
The Boole's Rule Integration, I_{BOOLE}, is:-

$$I_{BOOLE} = \frac{2}{45} h[7f(0,h) + 32f(1,h) + 12f(2,h) + 32f(3,h) + 7f(4,h)]$$
Equation 8.38

The I_{BOOLE} value is 0.943523628243005
PSD($\text{Sin}^{-1}(z)$,I_{BOOLE}) = -0.11096623924938

Rule A&S 25.4.15

m = 5

Abramowitz and Stegun Newton-Coates Numerical Quadrature Rule 25.4.15 is given (in two parts) by:-

$$p_1 = 19f(0,h) + 75f(1,h) + 50f(2,h)$$
Equation 8.39a

$$p_2 = 50f(3,h) + 75f(4,h) + 19f(5,h)]$$
Equation 8.39b

$$I_{15} = \frac{5}{288} h[p_1 + p_2]$$
Equation 8.40

The I_{15} value is 0.943135285501457
PSD($\text{Sin}^{-1}(z)$,I_{15}) = -0.069761794628532

Rule 18 A&S 24.4.18 p886

m = 8

In three parts, the primitive panel of Rule 18 is given by:-

$$p_1 = 989f(0,h) + 5888f(1,h) - 928f(2,h)$$
Equation 8.41a
$$p_2 = 10496f(3,h) - 4540f(4,h) + 10496(5,h)]$$
Equation 8.41b
$$p_3 = -928f(6,h) + 5888f(7,h) + 989f(8,h)]$$
Equation 8.41c

Note that some coefficients are negative.
Note also the striking symmetry of the coefficients within the
panel.

It follows that the first and often the only panel of Rule 18 is
given by:-

$$I_{RULE18} = \frac{4}{14175} h \left[\sum_{i=1}^{3} p_i \right]$$

Equation 8.42

The I_{RULE18} value is 0.942502651142021
$PSD(Sin^{-1}(z), I_{RULE18}) = -0.00263720431254$

So a single eight-interval application of Rule 18 yields a
good four-figure accuracy in the estimation of the Inverse Sine.

CHAPTER NINE
BASIC FUNCTIONS

PART ONE
LOGARITHMIC STUDIES

To summarise:-

$$\ln(\Phi^n) = n.\ln(\Phi) \approx 0.481211825059603 \times n$$
Equation 9.1

PART TWO
COMPLEX STUDIES

It is possible to relate π and Φ via the cosine of three-fifths π and the inverse cosine of half-Φ. It is convenient to mediate such activity by recourse to complex arithmetic involving $i = (-1)^{\frac{1}{2}}$, or in words the square root of minus one.[9.1]

We may commence by defining the local variables:-

$$z = \frac{\Phi}{2}$$

Equation 9.2

and:-

$$w = \frac{3\pi}{5}$$
Equation 9.3

Such that:-

$$\cos^{-1}(w) = -i.\ln\left(i\sqrt{1-w^2}\right) + w$$
Equation 9.4a

$$\cos^{-1}(z) = -i.\ln\left(i\sqrt{1-z^2}\right) + z$$
Equation 9.4b

and:-

$$\cos(w) = \frac{e^{iw} + e^{-iw}}{2}$$

Equation 9.5a

$$\cos(z) = \frac{e^{iz} + e^{-iz}}{2}$$

Equation 9.5b

Accordingly:-

$$\Lambda = \cos\left(\frac{3\pi}{5}\right)\cos^{-1}\left(\frac{\Phi}{2}\right) = \left(\frac{e^{iw} + e^{-iw}}{2}\right)\left(-i.\ln\left(i\sqrt{1 - z^2}\right) + z\right)$$

Equation 9.6

or:-

$$\Lambda = \frac{e^{iw}.\left(-i.\ln\left(i\sqrt{1 - z^2}\right) + z\right) + e^{iw}.\left(-i.\ln\left(i\sqrt{1 - z^2}\right) + z\right)}{2}$$

Equation 9.7

Therefore:-

$$\Lambda = \frac{1}{2}\cos(w)(\pi - 2\cos^{-1}(z)) = \frac{e^{iw} + e^{-iw}}{2}\left(-i.\ln\left(i\sqrt{1 - z^2}\right) + z\right)$$

Equation 9.8

and:-

$$\Lambda = \frac{e^{iw} + e^{-iw}}{2}.(-i)\ln\left(\sqrt{-i}.\sqrt{1 - z^2} + z\right)$$
$$= \frac{e^{iw} + e^{-iw}}{2}.(-i)\ln\left(\sqrt{z^2 - 1} + z\right)$$

Equation 9.9

so it follows that:-

$$\Lambda = \frac{\pi}{5}\left\{1 - \frac{\Phi + 2}{2[\Phi^2 - 2\Phi + 2]}\right\}$$

$$= \frac{\pi}{5}\left\{1 - \frac{\Phi + 2}{2[(\Phi - 1)^2 + 1]}\right\}$$

$$= \frac{\pi}{5}\left\{1 - \frac{\Phi + 2}{2[\Phi(\Phi - 2) + 2]}\right\}$$

$$= \frac{\pi}{5}\left\{1 - \frac{\Phi + 2}{2[1 + \phi^2]}\right\}$$

Equation 9.10

Because:-

$$\phi^2 + \phi = 1$$
Equation 9.11

it is self-evidently the case that for Natural Number j:-

$$(\phi^2 + \phi)^j = 1$$
Equation 9.12

and accordingly the following Binomial Expansions must hold:-

$$\phi^2 + \phi = 1$$
Equation 9.13a

$$\phi^4 + 2\phi^3 + \phi^2 = 1$$
Equation 9.13b

$$\phi^6 + 3\phi^5 + 3\phi^4 + \phi^3 = 1$$
Equation 9.13c

$$\phi^8 + 4\phi^7 + 6\phi^6 + 4\phi^5 + \phi^4 = 1$$
Equation 9.13d

So because for any Natural Numbers j and k:-

$$\phi^{2k} . \phi^{j-k} = \phi^{k+j}$$
Equation 9.14

it follows that:-

$$j = \sum_{k=0}^{j} \left[\frac{j!}{k!\,(j-k)!} \right] \phi^{2k} \cdot \phi^{j-k} = \sum_{k=0}^{j} \left[\frac{j!}{k!\,(j-k)!} \right] \phi^{k+j}$$
Equation 9.15

Now the Binomial Coefficient, ^nC_r, is given by:-

$$^n_rC = \binom{n}{k} = \frac{n!}{k!\,(n-k)!}$$
Equation 9.16

and accordingly:-

$$j = \sum_{k=0}^{j} \binom{j}{k} \phi^{k+j}$$
Equation 9.17

Two Expansions involving ϕ

It can be demonstrated that as n tends to infinity, j as defined above tends toward unity.
In practical terms:-

$$(\phi + \phi)^n = (2\phi)^n = 2^n \cdot \phi^n = \sum_{k=0}^{n} \binom{n}{k} \phi^n \to \infty$$
Equation 9.18

and it follows that:-

$$\sum_{k=0}^{n} \binom{n}{k} \phi^{k+n} \to 1$$
Equation 9.19

The Cumulative Distribution Function, CDF(ϕ)

The Cumulative Distribution Function, CDF(ϕ), is given by:-

$$1 = \sum_{k=0}^{n}\left[\frac{n!}{k!\,(n-k)!}\right]\phi^k.(1-\phi)^{n-k} = \sum_{k=0}^{n}\binom{n}{k}.\phi^k.(1-\phi)^{n-k}$$

Equation 9.20

Investigations of the Binomial Expansion Kernel (16 August 2022)

$$\Phi + 1 = \sum_{j=0}^{n}\frac{\phi^{2k}\phi^{j-k}}{\phi^k} = \sum_{j=0}^{n}\frac{\phi^{k+j}}{\phi^k} = \sum_{j=0}^{n}\phi^j$$

Equation 9.21

$$\Phi = \sum_{j=0}^{n}\phi^{2j}$$

Equation 9.22

$$0 = \phi^{k+j} = \phi^{2k}.\phi^{j-k}$$

Equation 9.23

Polynomials

Many small positive integers may be expressed as algebraic polynomials with small integer coefficients. For example:-

$$2\phi^4 + 3\phi^3 + 4\phi^2 + 4\phi + 2 = 7$$
Equation 9.24

and:-

$$2\phi^4 + 3\phi^3 + 4\phi^2 + 4\phi = 5$$
Equation 9.25

Clearly it follows that:-

$$(2\phi^4 + 3\phi^3 + 4\phi^2 + 4\phi + 2) + (2\phi^4 + 3\phi^3 + 4\phi^2 + 4\phi) = 12$$
Equation 9.26

and therefore that:-

$$4\phi^4 + 6\phi^3 + 8\phi^2 + 8\phi + 2 = 12$$
Equation 9.27

and:-

$$\frac{4\phi^4 + 6\phi^3 + 8\phi^2 + 8\phi + 2}{4} = 3$$
Equation 9.28

In addition to this we may observe from the definition of ϕ and Φ that:-

$$1 + \sqrt{5} = 2(1 + \phi) = 2\phi + 2$$
Equation 9.29

which implies that:-

$$(2\phi + 1)^2 = [2(1 + \phi) - 1]^2 = 4\phi^2 + 4\phi + 1 = 5$$
Equation 9.30

and because:-

$$\phi^2 + \phi = 1$$
Equation 9.31

we may write:-

$$4(\phi^2 + 1) = 4\phi^2 + 4 = 4$$
Equation 9.32

and hence:-

$$4\phi^2 + 4\phi + 5 = 9$$
Equation 9.33

PART FIVE
SPECIAL DERIVATION INVOLVING EULER'S IDENTITY

It is possible to show that for any Natural Number n:-

$$n = 2\phi^3 \left(n\phi + \frac{3}{2}n \right) = 2\phi^3 n \left(\phi + \frac{3}{2} \right)$$
Equation 9.34

from which at the expense of repetition we may clarify
that:-

$$2\phi^3 \left(n\phi + \frac{3}{2}n \right) = 2\phi^4 + 3\phi^3 = \phi^2 + \phi = 1$$
Equation 9.35

Using Euler's Identity

When i is the Square Root of Minus One, Euler's Identity is:-

$$e^{i\pi} + 1 = 0$$
Euler's Identity

and from the above it follows that:-

$$e^{i\pi} + 2\phi^4 + 3\phi^3 = 0$$
Equation 9.36

Allow that:-

$$\omega = \phi^2$$
Equation 9.37

Then we may erect the quadratic equation::-

$$2\omega^2 + 3\omega\phi + e^{i\pi} = 0$$
Equation 9.38

If locally a = 2, b = 3ϕ and c = $e^{i\pi}$ = -1, then the two roots of Equation 9.38 are given by:-

$$R1 = \frac{-b + \sqrt{b^2 - 4ac}}{2a}$$
Equation 9.39a

$$R1 = \frac{-b - \sqrt{b^2 - 4ac}}{2a}$$
Equation 9.39b

so that:-

$$R1 = \frac{-(3\phi) + \sqrt{(3\phi)^2 + 8 - 4ac}}{4} = \omega$$
Equation 9.40a

$$R2 = \frac{-(3\phi) - \sqrt{9\phi^2 + 8}}{2a} = -\left(\frac{\Phi}{2} + \frac{1}{2}\right)$$

Equation 9.40b

By application of Euler's Identity to $\phi^2 + \phi$ we may locally define a = 1, b = 1, c = $e^{i\pi}$ = -1 which yields:-

$$R1 = \frac{-(1) + \sqrt{(1)^2 + 4}}{2} = \phi$$

Equation 9.41a

$$R2 = \frac{-(1) - \sqrt{(1)^2 + 4}}{2} = -\Phi$$

Equation 9.41b

PART ONE
EXACT RELATIONS

<u>Exact Relations</u>

Allow that:-

$$z = \frac{\Phi}{2}$$
Equation 10.1

then:-

$$\chi = \tan^{-1}\left(\frac{\sqrt{1 - \left(\frac{\Phi}{2}\right)^2}}{\frac{\Phi}{2}}\right) = \tan^{-1}\left(\frac{\sqrt{\left(1 + \frac{\Phi}{2}\right)\left(1 - \frac{\Phi}{2}\right)}}{\frac{\Phi}{2}}\right)$$

$$= \tan^{-1}\left(\frac{2\sqrt{1 - \frac{\Phi^2}{4}}}{\Phi}\right)$$
Equation 10.2

hence:-

$$\chi = \tan^{-1}\left(\frac{\sqrt{4 - \Phi^2}}{\Phi}\right) = \tan^{-1}\left(\frac{\sqrt{1 - z^2}}{z}\right) = \frac{2}{\Phi}\sqrt{1 - \left(\frac{\Phi}{2}\right)^2} = \cos^{-1}(z)$$
Equation 10.3

PART TWO
APPROXIMATIONS

The Taylor Series Expansion for the Inverse Cosine (Φ/2)

Allow that:-

$$z = \frac{\Phi}{2}$$

Equation 10.4

and that:-

$$AC_{fid} = \cos^{-1}(z)$$
Equation 10.5

According to MathCad®.
Then as a Taylor Series Expansion:-

$$AC_{T1} = \frac{\pi}{2} - \sum_{n=0}^{m} \frac{(2n)!}{[(2^n).n!]^2} \times \frac{z^{2n+1}}{2n+1}$$
Equation 10.6

When m = 64, it is the case that PSD(AC_{fid}, AC_{T1}) = negative 0.000000000000265

The Taylor Series result is exact when m = ∞.

Approximations for the Inverse Cosine (of Φ/2 = z)

The Binomial Series

Allow that:-

$$BC(n,r) = {}^n_rC = \binom{n}{r} = \frac{n!}{r!\,(n-r)!}$$
Equation 10.7

then:-

$$AC_{BS} = \frac{\pi}{2} - \sum_{n=0}^{m} BC(2n,n) \cdot \frac{z^{2n+1}}{4^n \cdot (2n+1)}$$

Equation 10.8

When m = 64, PSD(AC_{fid}, AC_{BS}) = -0.000000000000265
So from an operational point of view, the outcome of the Binomial Series is identical to that of the Taylor Series Expansion.

Approximations for the Inverse Tangent (of $\Phi/2 = z$)

The MathCad® function:-

$$atan\left(\frac{\Phi}{2}\right) = 0.680214976880921$$

Equation 10.9

is assumed exact and therefore fiducial.

NXP Semiconductors AN5016 Approximation

$$ATANapp = \frac{\frac{\Phi}{2}\left(1 + \frac{4}{15} \cdot \left(\frac{\Phi}{2}\right)^2\right)}{1 + \frac{3}{5} \cdot \left(\frac{\Phi}{2}\right)^2}$$

Equation 10.10

For this NXP Semiconductors expression the computed Percentage Specific Defect PSD(atan($\Phi/2$),ATANapp) = negative 0.304051078621919 which of course represents an accuracy better than one part in three hundred.

Second Order Rational Approximation[10.1]

$$atanapp_{Benammar} = \frac{\frac{\Phi}{2}}{1 + 0.0443\frac{\Phi}{2} + 0.2310\left(\frac{\Phi}{2}\right)^2}$$

Equation 10.11

PSD(atan(Φ/2),atanapp$_{Benammar}$) is -0.195778623859442

The Second Euler Series

$$AT_{E2} = \sum_{n=0}^{m} \frac{2^{2n}.(n!)^2}{(2n+1)!} \times \frac{\left(\frac{\Phi}{2}\right)^{2n+1}}{\left(1+\left(\frac{\Phi}{2}\right)^2\right)^{n+1}}$$

Equation 10.12

When m = 64 the Percentage Standard Deviation PSD(atan(Φ/2),AT$_{E2}$) is 0.000000000000016

Selected Approximations for π

The First Nilakantha Formula

This approximant may date from the thirteenth century AD and is given by:-

$$AP_{N1} = 4\left[\frac{3}{4} + \sum_{n=0}^{m} \frac{(-1)^n}{(2n+3)^3 - (2n+3)}\right] \approx 3.14159352291775$$

Equation 10.13

m = 64 and PSD(π,AP$_{N1}$) = -0.000027671568268

The Second Nilakantha Formula

Also of thirteenth century vintage this is:-

$$AP_{N2} = \sum_{n=0}^{m} \frac{(5n+3).n!.(2n)!}{2^{n+1}.(3n+2)!} \approx 3.14159265358979$$
Equation 10.14

m = 12 and PSD(π,AP$_{N2}$) = 0.000000000000042

The First Adegoki-Layeni Series (2016) (modified by J Warren)

One of the better recent offerings, but not the equal of Ramanujan II:-

$$AP_{AL1} = \sqrt{3}.\frac{3}{8}.\sum_{n=0}^{m} \frac{(-1)^n}{8}\left[\frac{4}{3n+1} + \frac{2}{3n+2}\right]$$
Equation 10.15

m = 64 and PSD(π,AP$_{AL1}$) = -0.000000000000014

<u>The Simplification of the Taylor Series Expansion for the Inverse Cosine (Φ/2)</u>

This simplification assumes that n = 5, and that accordingly n! = 120, (n!)^2 = 14400, (2n)! = 3628800, (2n).n! = 3840 and [(2n).n!]^2 = 14755600

Under this condition:-

$$\left(\frac{\Phi}{2}\right)^{2n+1} = \frac{\Phi^{11}}{2^{11}}$$
Equation 10.16

therefore:-

$$\frac{(2n)!}{[2^n.n!]^2} \times z^{2n+1} = \frac{(2n)!}{[2^n.n!]^2} \times \frac{\Phi^{11}}{2^{11}}$$
Equation 10.17

which is the same as:-

$$\frac{\prod_{j=n+1}^{2n} j}{2^{2n}.n!} \times \frac{\Phi^{2n+1}}{2^{2n+1}} = \frac{\prod_{j=n+1}^{2n} j}{n!} \times \frac{\Phi^{2n+1}}{2^{4n+1}}$$
Equation 10.18

From the above it is then possible to define the Simplified Taylor Series for z, AC_{T1}, as:-

$$AC_{T1} = -z + \frac{\pi}{2} - \sum_{n=1}^{m} \frac{\prod_{j=n+1}^{2n} j}{2^{2n}.n!} \times \frac{z^{2n+1}}{2n+1}$$
Equation 10.19

or:-

$$AC_{T1} = -z + \frac{\pi}{2} - \sum_{n=1}^{m} \prod_{j=n+1}^{2n} j \times \frac{z^{2n+1}}{(2^{2n}.n!)(2n+1)}$$
Equation 10.20

In both versions m =70 and PSD(AC_{fid},AC_{T1}) = negative 0.000000000000018

Further simplifications allow us to write:-

$$AC_{T1} = -z + \frac{\pi}{2} - \sum_{n=1}^{m} \frac{\prod_{j=n+1}^{2n} j}{\prod_{k=1}^{n} k} \times \frac{\Phi^{2n+1}}{2^{4n+1}.(2n+1)}$$
Equation 10.21

whose accuracy is identical to the previous two forms, though I have not conducted experiments to compare the efficiency of any of the three computations.

The Simplification of the Binomial Series for the Inverse Cosine (Φ/2)

As per Equation 10.8:-

$$ AC_{BS} = \frac{\pi}{2} - \sum_{n=0}^{m} BC(2n, n) . \frac{z^{2n+1}}{4^n . (2n+1)} $$

Equation 10.8

yields for m = 70 a PSD(acos(Φ/2),AC$_{BS}$) = negative 0.000000000000071

Simplifications allow is to write:-

$$ AC_{BS} = -z + \frac{\pi}{2} - \sum_{n=1}^{m} \frac{\prod_{j=n+1}^{2n} j}{\prod_{k=1}^{n} k} \times \frac{\Phi^{2n+1}}{2^{4n+1} . (2n+1)} $$

Equation 10.21a

that is identical to the Taylor Series simplification albeit the accuracy is a trifle worse. Having said which the PSD(π/5,AC$_{BS}$) is zero.

Thus by equivalation:-

$$ \frac{\pi}{5} = -z + \frac{\pi}{2} - \sum_{n=1}^{m} \frac{\prod_{j=n+1}^{2n} j}{\prod_{k=1}^{n} k} \times \frac{\Phi^{2n+1}}{2^{4n+1} . (2n+1)} $$

Equation 10.22

so:-

$$ z - \frac{\pi}{5} - \frac{\pi}{2} = -\sum_{n=1}^{m} \frac{\prod_{j=n+1}^{2n} j}{\prod_{k=1}^{n} k} \times \frac{\Phi^{2n+1}}{2^{4n+1} . (2n+1)} $$

Equation 10.23

or by alternating signs:-

$$\frac{\pi}{2} - \frac{\pi}{5} - z = \sum_{n=1}^{m} \frac{\prod_{j=n+1}^{2n} j}{\prod_{k=1}^{n} k} \times \frac{\Phi^{2n+1}}{2^{4n+1}.(2n+1)}$$

Equation 10.24

from which:-

$$\frac{3}{10}\pi = \frac{\Phi}{2} + \sum_{n=1}^{m} \frac{\prod_{j=n+1}^{2n} j}{\prod_{k=1}^{n} k} \times \frac{\Phi^{2n+1}}{2^{4n+1}.(2n+1)}$$

Equation 10.25

So that by isolation of π on the LHS we attain the key relation:-

$$\pi = \frac{10}{3}\left[\frac{\Phi}{2} + \sum_{n=1}^{m} \frac{\prod_{j=n+1}^{2n} j}{\prod_{k=1}^{n} k} \times \frac{\Phi^{2n+1}}{2^{4n+1}.(2n+1)}\right]$$

Equation 10.26

Using this approximant for m = 70 and Pi $\equiv$ RHS the relevant PSD(π,pi) is zero.

PART THREE
FURTHER IDENTITIES AND APPROXIMATIONS

Without undue pre-amble we recall that:

$$\Phi = \frac{1 + \sqrt{5}}{2} = 1.61803398874989$$
Equation 10.27

and that:-

$$\phi = \Phi - 1 = 0.618033988749895$$
Equation 10.28

So:-

$$\Phi = \frac{1 + \sqrt[4]{25}}{2}$$
$$= \frac{1 + \sqrt[4]{2^4 + 3^2}}{2}$$
$$= \frac{1 + \sqrt[4]{2^4 + 3^2}}{\sqrt[4]{16}}$$
$$= \frac{1}{\sqrt[4]{16}} + \frac{\sqrt[4]{2^4 + 3^2}}{\sqrt[4]{16}}$$
Equation 10.29

or in approximate terms:-

$$\Phi_{approx} = \left[\sqrt[3]{\frac{1}{4}\left(2 + \frac{2^4 + 3^2}{4}\right)} \right]^2 = \left[\frac{1}{4}\left(2 + \frac{2^4 + 3^2}{4}\right)\right]^{\frac{2}{3}}$$
Equation 10.30

The Percentage Specific Defect (Φ,Φ_{approx}) is:-

$$PSD\left(\Phi, \Phi_{approx}\right) = PSD\left(\frac{1+\sqrt{5}}{2}, \left[\frac{1}{4}\left(2 + \frac{2^4 + 3^2}{4}\right)\right]^{\frac{2}{3}}\right)$$

$$= -0.140171591154085$$

Equation 10.31

We shall meet this defect, or its positive or negative near-equivalents, on numerous succeeding occasions, even in the context of mineral crystallography.

Pi and the Square Root of the Major Phidian Ratio

Note these roots:-

$\sqrt{2} = 1.4142135623631$	**Equation 10.32a**
$\sqrt{3} = 1.73205080756888$	**Equation 10.32b**
$\sqrt{5} = 2.23606797749979$	**Equation 10.32c**

By way of approximation we may write:-

$$\pi \approx \sqrt{2} + \sqrt{3} = 3.14626436994197$$

Equation 10.33

The Percentage Specific Defect PSD between π and $2^{\frac{1}{2}} + 3^{\frac{1}{2}}$ is given by:-

$$PSD\left(\pi, \sqrt{2} + \sqrt{3}\right) = -0.148705350034519$$

Equation 10.34

and exactly:-

$$\sqrt{5} = \sqrt[4]{2^4 + 3^2}$$

Equation 10.35

whilst also:-

$$\sqrt{\Phi} = \sqrt{\frac{1^2}{2} + \left(\frac{\sqrt{5}}{2}\right)^1} \approx \sqrt[3]{\frac{1}{2} + \frac{2^4 + 3^2}{16}} = \sqrt[3]{\frac{1}{4}\left(2 + \frac{2^4 + 3^2}{4}\right)} = \sqrt[3]{\frac{33}{16}}$$

Equation 10.36

from which:-

$$\frac{33}{16} = \frac{1}{2} + \frac{2^4 + 3^2}{16} = 2.0625$$

Equation 10.37

The relevant PSD is:-

$$PSD\left(\sqrt{\Phi}, \sqrt[3]{\frac{1}{4}\left(2 + \frac{2^4 + 3^2}{4}\right)}\right) = -0.070061252681415$$

Equation 10.38

We were able to write:-

$$\sqrt{\Phi} \approx \sqrt[3]{\frac{33}{16}}$$

Equation 10.39

so the PSD of this expression is given by:-

$$PSD\left(\sqrt{\Phi}, \sqrt[3]{\frac{33}{16}}\right) = -0.070061252681415$$

Equation 10.40

Noting that:-

$$A = \sqrt{1^2 + 1^2 + 1^2} = \sqrt{3} \qquad \textbf{Equation 10.41a}$$
$$B = \sqrt{2^2 + 2^2 + 1^2} = 3 \qquad \textbf{Equation 10.41b}$$

$$C = \sqrt{1^2 + 2^2 + 1^2} = \sqrt{6} \qquad \textbf{Equation 10.41c}$$

a is the Approximation of π (Equation 10.33) which is $2^{\frac{1}{2}}+3^{\frac{1}{2}}$ so we can move forward with:-

$$\frac{25}{a} \approx 5 + \left[\frac{1}{4}\left(2 + \frac{2^4 + 3^2}{4}\right)\right]^{\frac{3}{2}}$$

$$\textbf{Equation 10.42}$$

for which the PSD is:-

$$PSD\left(\frac{25}{a}, 5 + \left[\frac{1}{4}\left(2 + \frac{2^4 + 3^2}{4}\right)\right]^{\frac{3}{2}}\right) = -0.202732498725163$$

$$\textbf{Equation 10.43}$$

Re-arrangement of Equation 10.42 yields:-

$$a \approx \frac{25}{5 + \left[\frac{1}{4}\left(2 + \frac{2^4 + 3^2}{4}\right)\right]^{\frac{3}{2}}}$$

$$\textbf{Equation 10.44}$$

from which, with appropriate substitution for the square-bracted expression, we attain:-

$$a \approx \frac{25}{5 + \left(\frac{33}{16}\right)^{\frac{3}{2}}}$$

$$\textbf{Equation 10.45}$$

for which the PSD is given by:-

$$PSD\left(a, \frac{25}{5 + \left(\frac{33}{16}\right)^{\frac{3}{2}}}\right) = -0.202322325618956$$

Equation 10.46

But when we essay the formula:-

$$\pi \approx \frac{25}{5 + \left(\frac{33}{16}\right)^{\frac{3}{2}}}$$

Equation 10.47

we obtain:-

$$PSD\left(\pi, \frac{25}{5 + \left(\frac{33}{16}\right)^{\frac{3}{2}}}\right) = 0.053917839706947$$

Equation 10.48

an essentially four-fold improvement in the absolute accuracy.

If we now square both sides of Equation 10.39 we get:-

$$\Phi \approx \left(\frac{33}{16}\right)^{\frac{2}{3}}$$

Equation 10.49

With respect to the exponentiated fraction in the RHS denominator of Equation 10.49 we may expand that fraction as:-

$$\left(\frac{33}{16}\right)^{\frac{3}{2}} = \left(\frac{33}{16}\right)^{\frac{2}{3}} \times \left(\frac{33}{16}\right)^{\frac{5}{6}}$$

Equation 10.50

So Equation 10.47 expands to:-

$$\pi \approx \frac{25}{5 + \left(\frac{33}{16}\right)^{\frac{2}{3}}\left(\frac{33}{16}\right)^{\frac{5}{6}}} \approx \frac{25}{5 + \Phi\left(\frac{33}{16}\right)^{\frac{5}{6}}}$$

Equation 10.51

$$PSD\left(\pi, \frac{25}{5 + \Phi\left(\frac{33}{16}\right)^{\frac{5}{6}}}\right) = 0.001845123340767$$

Equation 10.52

Factorisations yield the form:-

$$\pi \approx \frac{25}{5 + \Phi\left(\frac{3.11}{2^4}\right)^{\frac{5}{6}}}$$

Equation 10.53

for which the unchanged Percentage Specific Defect is:-

$$PSD\left(\pi, \frac{25}{5 + \Phi\left(\frac{3.11}{2^4}\right)^{\frac{5}{6}}}\right) = 0.001845123340767$$

Equation 10.54

The defect is now better than one part in 54196: What old-fashioned scientists like me would have called "quantitative" in the dreadful jargon of the Sixties.

Allow that, locally, and for the purposes of simplification:-

$$z = \left(\frac{3.11}{2^4}\right)^{\frac{5}{6}}$$

Equation 10.55

such that:-

$$\pi \approx \frac{25}{5 + \Phi z}$$
Equation 10.56

then:-

$$z = \left(\frac{3.11}{2^4}\right)^{\frac{5}{6}} = \frac{2^{\frac{2}{3}}.33^{\frac{5}{6}}}{2^4} = \frac{33^{\frac{5}{6}}.e^0}{8.\sqrt[3]{2}} = \frac{33^{\frac{5}{6}}}{8.\sqrt[3]{2}}$$
Equation 10.57

Due to numerical error the relevant PSD is:-

$$PSD\left(\pi, \frac{25}{5 + \Phi z}\right) = 0.001845123340781$$
Equation 10.58

Furthermore:-

$$\frac{25}{a} - 5 \approx \left[\frac{1}{4}\left(2 + \frac{2^4 + 3^2}{4}\right)\right]^{\frac{3}{2}}$$
Equation 10.59

so also that numerically:-

$$\sqrt[3]{\frac{1}{4}\left(2 + \frac{2^4 + 3^2}{4}\right)} = \sqrt[3]{\frac{1}{4}\left(2 + \frac{2^4 + 3^2}{4}\right)}$$
$$= \sqrt[3]{\frac{1}{4}\left(2 + \frac{25}{4}\right)}$$
$$= \sqrt[3]{\frac{1}{4}(8.25)}$$
$$= \sqrt{\frac{33}{16}}$$

$$= \sqrt{\Phi}$$
Equation 10.60

From which it follows by squaring both sides that:-

$$\Phi = \frac{33^{\frac{2}{3}}}{16}$$
Equation 10.61

at this stage:-

$$PSD\left(\Phi, \frac{33^{\frac{2}{3}}}{16}\right) = -0.140171591154085$$
Equation 10.62

We have seen from Equations 10.56 and 10.58 that π is very near indeed to $25/(5+\Phi z)$. Pursuing this fact we may write:-

$$\frac{25}{\pi} \approx 5 + \Phi z$$
Equation 10.63

and that therefore:-

$$\Phi \approx \frac{1}{z}\left(\frac{25}{\pi} - 5\right)$$
Equation 10.64

so:-

$$\Phi \approx \frac{1}{\frac{33^{\frac{5}{6}}}{8 \cdot \sqrt[3]{2}}}\left(\frac{25}{\pi} - 5\right) = \frac{8 \times \sqrt[3]{2}}{33^{\frac{5}{6}}}\left(\frac{25}{\pi} - 5\right)$$
Equation 10.65

A little further wrangling allows us to define an approximation of the Major Ratio, Φ_{approx}, in terms of:-

$$\Phi_{approx} \approx \frac{2^{\frac{10}{3}}.5}{33^{\frac{5}{6}}}\left(\frac{5}{\pi}-1\right) = 1.61795366784591$$

Equation 10.66

For this expression the Percentage Specific Defect is given by:-

$$PSD(\Phi, \Phi_{approx}) = 0.004964104867027$$

Equation 10.67

This is a shortfall of one part in 20144.6187537714 or in other words in part in 20144+φ.

CHAPTER ELEVEN
SOME SPACIAL RELATIONS

Huntley's Cone[11.1,11.2]

Huntley's Cone[11.1] was introduced by the late Professor Herbert Edwin Huntley as an exercise in his brief but important 1970 book "The Divine Proportion".

Huntley stated the problem as:-

> *"The circumference of the base of a right circular cone,*
> *of which the semi-vertical angle is 54 degrees and the slant side*
> *measures one foot, is πΦ feet, and the curved surface is ½πΦ*
> *square feet.*
> *The proof of this, with the help of the table on p.40,*
> *is left to the reader."*

(The foot is an obsolete British Imperial measure of length of approximately twenty-five centimeters).

I did not consult the table on Page 40, but nevertheless attempted, if not a proof, then a demonstration in the following terms.

The highly oblate form of the Huntley Cone is illustrated by the GeoGebra® plot offered as Figure 11.1

Allow that l (lower-case letter el) is the Slant Side Length as stated, and also that Basal Radius, r, is:-

$$r = \frac{c}{2\pi}$$
Equation 11.1

where c is the Basal Circumference, stated by Huntley to be πΦ:-
Then by substitution for c:-

$$r = \frac{\pi\Phi}{2\pi}$$
Equation 11.2

from which it follows that Radius r is given by:-

$$r = \frac{\Phi}{2} = 0.809016994374947$$

Equation 11.3

Having computed the Radius it is now mete to compute the Cone Height, h, using Pythagoras:-

$$h = \sqrt{l^2 + r^2} = 0.587785252292473$$
Equation 11.4

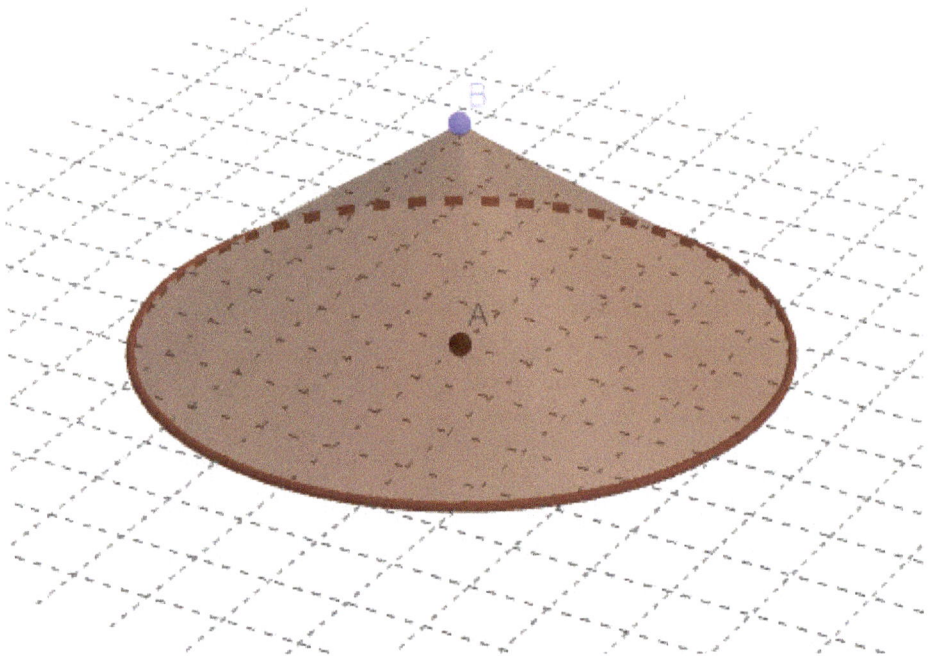

Figure 11.1
The Huntley Cone

The curved area of the cone surface above the flat Base is called its Nappe. Thus the Area of the Nappe is given by:-

$$A_{nappe} = \pi r l = \pi \frac{\Phi}{2} \times 1$$
Equation 11.5

which agrees with the area of Huntley's "curved surface":-

$$A_{given} = \frac{1}{2}\pi\Phi$$
Equation 11.6

This completes the demonstration.

Knowledge of angles is not required.

Notwithstanding which, I attempt to identify and compute available angles in the following terms:-

At first I was not sure what Huntley meant by "semi-vertical angle".

But it seemed to me that there were only two available angles in a right cone:-

(1)   Basal Angle, β

$$\beta = \tan^{-1}\left(\frac{h}{r}\right) = 0.628318530717959 = 36°$$
Equation 11.7

(2)   Semi-Apical Angle, α

$$\alpha = \tan^{-1}\left(\frac{r}{h}\right) = 0.942477796076938 = 54°$$
Equation 11.8

which may, for the given case, be simplified in the following terms:-

$$\alpha = \tan^{-1}\left(\frac{r}{h}\right) = \tan^{-1}\left(\frac{\frac{\Phi}{2}}{\sqrt{1^2 - \left(\frac{\Phi}{2}\right)^2}}\right) = \tan^{-1}\left(\frac{\Phi}{\sqrt{4 - \Phi^2}}\right)$$

Equation 11.9

from which it follows that the Full Apical Angle, γ, is[11.3]:-

$$\gamma = 2.\tan^{-1}\left(\frac{r}{h}\right) = 1.88495559215388 = 108°$$

Equation 11.10

The Surface Area and the Volume of the Huntley Cone

The Surface Area, A, is the sum of the Basal Area, A_{base}, and the Nappe Area, A_{nappe}, which is given by:-

$$A = A_{base} + A_{nappe} = \pi r^2 + \pi r l = \pi r(r + l) = 4.59780093263389$$

Equation 11.11

Meanwhile the Cone Volume, V, is given by:-

$$V = \frac{1}{3}.\pi r^2.h = 0.402867832936001$$

Equation 11.12

Accordingly the Ratio of Cone Surface Area to Volume, R_{cone}, is:-

$$R_{cone} = \frac{A}{V} = \frac{\pi r(r + l)}{\frac{1}{3}.\pi r^2.h} = \frac{3(r + l)}{r.h} = 11.4126781955418$$

Equation 11.13

The Maximal Surface Area of a Cylinder

It is possible to show that the surface area of a cylinder of given Radius (or height) is a maximum when its Ratio of Radius to Height has the value of Φ. Figure 11.2 is a GeoGebra® plot illustrative of the situation.

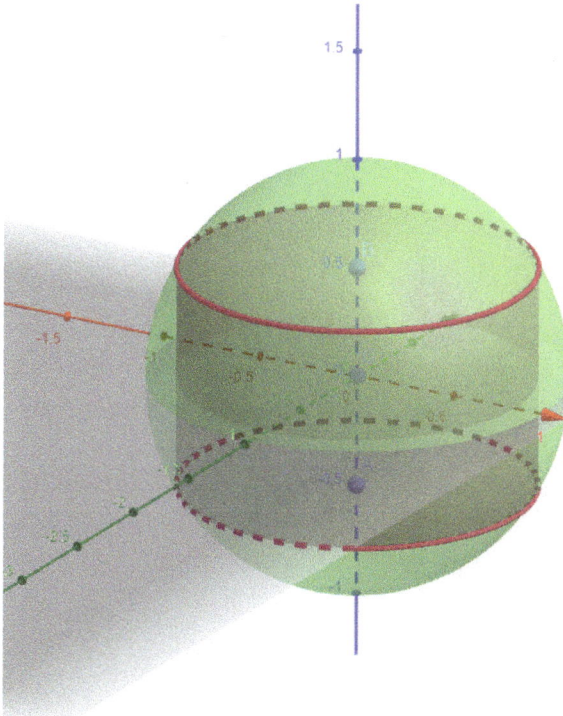

Figure 11.2
The Maximal Surface Cylinder and its
Circumscriptive Sphere

The red figure represents the maximal cylinder whose diameter is Φ and whose Height is unity. The green figure is the Circumscriptive Circle whose Radius, R, is (in native metrics):-

$$R = \sqrt{\left(\frac{\Phi}{2}\right)^2 + \left(\frac{1}{2}\right)^2}$$

Equation 11.14

Figure 11.3 is a two-dimensional section through the common center (i.e. the (x-z plane).

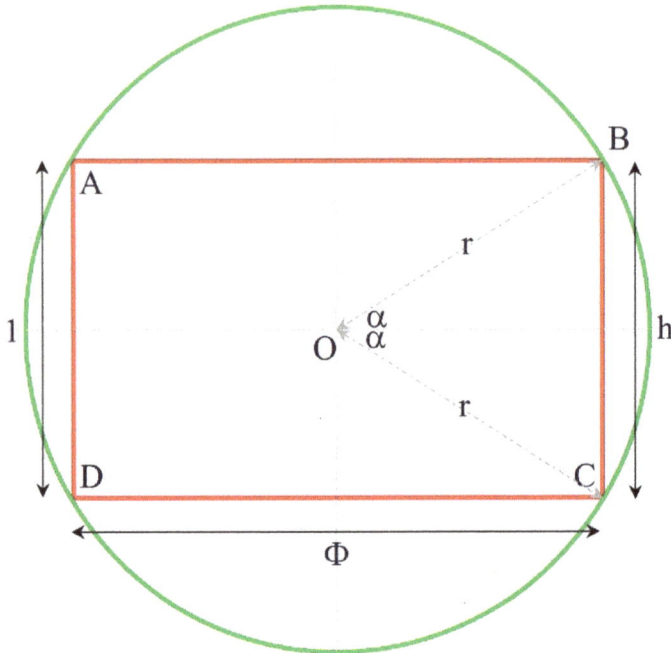

Figure 11.3
Annotated (x,z) Section of the Maximal Cylinder
and its Circumscriptive Sphere

The Reference Sphere

For ratios' assessments purposes it is convenient to define sphere formulae and angular datums for later analyses, especially analyses involving discrete series.

Therefore r_{sphere}, a common working Radius of the Sphere and its circumscribed cylinder will provisionally be set to unity. Also the Volume of a Sphere, V_{sphere} is given by:-

$$V_{sphere} = \frac{4}{3}.\pi.r_{sphere}^{3}$$

Equation 11.15

and the Sphere Surface Area S_{sphere} is:-

$$S_{sphere} = 4\pi.r_{sphere}$$

Equation 11.16

With regard to datum angles at or about the loci of interest we may define the Angle α_{center} between the figures' center O and the Phidian Rectangle north-east corner B as:-

$$\alpha_{center} = \tan^{-1}\left(\frac{\frac{1}{2}}{\frac{\Phi}{2}}\right) = \tan^{-1}\left(\frac{1}{\Phi}\right) = 0.553574358897045$$

Equation 11.17

or in degrees:-

$$\alpha_{center} = 180.\frac{\alpha_{center}}{\pi} = 31.717474411461°$$

Equation 11.18

Empirical (Discrete Series) Treatment

Values of the geometrical parameters of the rectangle between Sweep Angles 0 and $\pi/2$ of the north-east sector were computed in n = 16 equal intervals. Sweeps for n = 32 on n = 64 were essayed but found to confer no greater convenience or precision.

Specifically, the Angular Increment $\Delta\alpha$ was defined to be:-

$$\Delta\alpha = \frac{1}{4} \cdot \frac{1}{n}$$
Equation 11.19

and the Lower Bound Datum Angle was defined as:-

$$\alpha_{datum} = \alpha_{center} - \frac{n}{2} \cdot \Delta\alpha$$
Equation 11.20

Accordingly, Lapse Angle α_i for Interval i was:-

$$\alpha_i = \alpha_{datum} + i.\Delta\alpha$$
Equation 11.21

and the Cartesian Co-ordinates (x_i, y_i) of the Swept Corners were accordingly:-

$$x_i = r_i.\cos(\alpha_i) \qquad\qquad y_i = r_i.\sin(\alpha_i)$$
Equation 11.22a   **Equation 11.22b**

All r_i were taken as unity.
The Surface Area of the ith. Cylinder CS_i is therefore:-

$$CS_i = 2\pi x_i(x_i + 2y_i) = 2\pi x_i(x_i + h_i)$$
Equation 11.23

Equation 11.23 defines a hyperbolic paraboloid.
The corresponding Volume is:-

$$CV_i = \pi x_i^2 y_i = \pi x_i^2 \frac{h_i}{2}$$
Equation 11.24

In addition we may define the Ratios:-

$$R_i = \frac{x_i}{y_i} = \frac{2x_i}{h_i} = \frac{1}{\tan(\alpha_i)} = \cot(\alpha_i)$$
Equation 11.25

and:-

$$T_i = \frac{CS_i}{CV_i} = \frac{2\pi x_i(x_i + 2y_i)}{\pi x_i{}^2 y_i} = \frac{2\pi x_i{}^2 + 4\pi x_i y_i}{\pi x_i{}^2 y_i} = 2\left(\frac{1}{y_i} + \frac{2}{x_i}\right)$$

Equation 11.26

Expressed in terms of the Angle a_i the Cylinder Surface Area CSa_i is:-

$$CS\alpha_i = 2\pi x_i\left[x_i + r_i.\frac{\sin(2\alpha_i)}{\sin\left(\dfrac{\pi - 2\alpha_i}{2}\right)}\right]$$

Equation 11.27

Analytical (Native Metrics) Treatment

Analysis of the geometry yields results very similar to the foregoing except that rather than the arbitrary definition of Radius as unity, we may more rationally use the Pythagorean Theorem to define it as:-

$$r_\Phi = \sqrt{\left(\frac{\Phi}{2}\right)^2 + \left(\frac{1}{2}\right)^2}$$

Equation 11.28

whilst the Central Angle becomes α_Φ given by:-

$$\alpha_\Phi = \tan^{-1}\left(\frac{1}{\alpha_\Phi}\right)$$

Equation 11.29

From these facts it follows that:-

$$x_\Phi = r_\Phi.\cos(\alpha_\Phi)$$
Equation 11.30a

$$y_\Phi = r_\Phi.\sin(\alpha_\Phi)$$
Equation 11.30b

And also it follows that the analytic Cylinder Surface Area CS_α is given by:-

$$R_\Phi = \frac{x_\Phi}{y_\Phi} = \cot(\alpha_\Phi)$$

Equation 11.31

The analytic Cylinder Surface Area CS_Φ is accordingly:-

$$CS_\Phi = 2\pi x_\Phi (x_\Phi + 2y_\Phi) = 9.19560186526779$$

Equation 11.32

Equation 11.32 simplifies in the terms below:-

$$
\begin{aligned}
CS_\Phi &= 2\pi x_\Phi (x_\Phi + 2y_\Phi) \\
&= 2\pi r_\Phi.\cos(\alpha_\Phi)\left[r_\Phi.\cos(\alpha_\Phi) + 2r_\Phi.\sin(\alpha_\Phi)\right] \\
&= 2\pi r_\Phi{}^2.\cos(\alpha_\Phi)\left[\cos(\alpha_\Phi) + 2.\sin(\alpha_\Phi)\right] \\
&= 2\pi r_\Phi{}^2.\{\cos(\alpha_\Phi)^2 + 2.\sin(\alpha_\Phi).\cos(\alpha_\Phi)\} \\
&= 2\pi r_\Phi{}^2.\{\cos(\alpha_\Phi)^2 + \sin(2\alpha_\Phi)\}
\end{aligned}
$$

Equation 11.33

Now it is the case that:-

$$\cos(\alpha_\Phi)^2 = \frac{\frac{\Phi}{2}}{\sqrt{\left(\frac{\Phi}{2}\right)^2 + \left(\frac{1}{2}\right)^2}} \times \frac{\frac{\Phi}{2}}{\sqrt{\left(\frac{\Phi}{2}\right)^2 + \left(\frac{1}{2}\right)^2}} = \frac{\frac{\Phi^2}{4}}{\left(\frac{\Phi}{2}\right)^2 + \left(\frac{1}{2}\right)^2}$$

Equation 11.34

and that:-

$$\sin(2\alpha_\Phi) = 2\frac{\frac{\Phi}{2}}{\sqrt{\left(\frac{\Phi}{2}\right)^2 + \left(\frac{1}{2}\right)^2}}\frac{\frac{\Phi}{2}}{\sqrt{\left(\frac{\Phi}{2}\right)^2 + \left(\frac{1}{2}\right)^2}} = \frac{\frac{2\Phi}{4}}{\left(\frac{\Phi}{2}\right)^2 + \left(\frac{1}{2}\right)^2}$$

Equation 11.35

Therefore we may re-quote the Phidian Cylinder Surface Area CS_Φ as:-

$$CS_\Phi = 2\pi r_\Phi{}^2.\{\cos(\alpha_\Phi)^2 + \sin(2\alpha_\Phi)\}$$
Equation 11.36

$$CS_\Phi = 2\pi r_\Phi{}^2.\left\{\frac{\frac{\Phi^2}{4}}{\left(\frac{\Phi}{2}\right)^2 + \left(\frac{1}{2}\right)^2} + \frac{\frac{2\Phi}{4}}{\left(\frac{\Phi}{2}\right)^2 + \left(\frac{1}{2}\right)^2}\right\}$$

$$CS_\Phi = 2\pi r_\Phi{}^2.\left\{\frac{\frac{\Phi^2 + 2\Phi}{4}}{\left(\frac{\Phi}{2}\right)^2 + \left(\frac{1}{2}\right)^2}\right\}$$

$$CS_\Phi = 2\pi r_\Phi{}^2.\left\{\frac{1}{4}.\frac{\Phi^2 + 2\Phi}{\left(\frac{\Phi}{2}\right)^2 + \left(\frac{1}{2}\right)^2}\right\}$$

$$CS_\Phi = 2\pi.\left[\left(\frac{\Phi}{2}\right)^2 + \left(\frac{1}{2}\right)^2\right]\left\{\frac{1}{4}.\frac{\Phi^2 + 2\Phi}{\left(\frac{\Phi}{2}\right)^2 + \left(\frac{1}{2}\right)^2}\right\}$$
Equation 11.37

and by further reduction:-

$$CS_\Phi = 2\pi\left(\frac{1}{4}.\frac{\Phi^2 + 2\Phi}{1}\right) = \frac{\pi(\Phi^2 + 2\Phi)}{2}$$
Equation 11.38

The Surface Area of the Circumscriptive Sphere SS_Φ is given by:-

$$SS_\Phi = 4\pi.r_\Phi$$
Equation 11.39

So because:-

$$\sqrt{\left(\frac{\Phi}{2}\right)^2 + \left(\frac{1}{2}\right)^2} = \frac{\sqrt{\Phi^2 + 1}}{2}$$

Equation 11.40

the Surface Area Ratio of the Phidian Cylinder to its Circumscriptive Sphere T_Φ is given by:-

$$T_\Phi = \frac{CS_\Phi}{SS_\Phi} = \frac{\pi \dfrac{\Phi^2 + 1}{2}}{4\pi.r_\Phi} = \frac{\pi \dfrac{\Phi^2 + 1}{2}}{4\pi.\sqrt{\left(\frac{\Phi}{2}\right)^2 + \left(\frac{1}{2}\right)^2}} = 0.769420884293813$$

Equation 11.41

Cancelling π and further simplification then enables us to write:-

$$T_\Phi = \frac{\Phi^2 + 1}{8.\sqrt{\left(\frac{\Phi}{2}\right)^2 + \left(\frac{1}{2}\right)^2}} = \frac{1 + 3\Phi}{4\sqrt{\Phi + 2}}$$

Equation 11.42

Numerical Results

Table 11.1 summarises the salient input data and results for the Phidian Maximal Surface Area Cylinder.
Table 11.2 shows the sixteen interval EXCEL® elaboration for inscribed rectangles of different aspect.

The Major Ratio of Phidias	Φ	1.618033988749890

THE STANDARD SPHERE

Radius	r_{sphere}	0.951056516
Surface Area	S_{sphere}	3.603359441571710
Volume	V_{sphere}	11.951328658966200

INCYLINDER PARAMETERS

Number of Intervals	n	16
Central Angle of Seriation (radians)	α_{center}	0.553574358897045
Central Angle of Seriation (degrees)	αdeg_{center}	31.717474411461000
Angular Increment	$\Delta\alpha$	0.015625000000000
Series Datum Angle	α_{datum}	0.428574358897045
Central Radius of Cylinder	r_i	0.951056516

LOCATION OF MAXIMAL CYLINDER SURFACE

Trigonometric Ratio Value for Φ	$(x_i/y_i)_{n/2}$	1.618033988749900
CoTangent of Central Angle α_Φ	$cot(\alpha_\Phi)$	1.618033988749890
Fractional Defect	FD(F,cot(aF))	0
Cylinder Surface at α_Φ	$CS_{\alpha\Phi}$	9.195601865267790
Cylinder/Sphere Surface Ratio at α_Φ	$(CS/S_{sphere})_{\alpha\Phi}$	2.551952425056120

Table 11.1
Summary for the Phidian Maximal Surface Cylinder

Results in bold red are maxima and results in bold green are optima.

It is now manifest that:-

$$\Phi = R_{\frac{n}{2}} = \frac{x_{\frac{n}{2}}}{y_{\frac{n}{2}}} = \cot\left(\alpha_{\frac{n}{2}}\right)$$

Equation 11.43

Serial i	Angle α_i	x_i	y_i	Ratio x_i/y_i	Cylinder Surface CS_i	Cylinder Volume CV_i	Surface Ratio CS_i/S_{sphere}	Volume Ratio CV_i/V_{sphere}	Fractional Defect $FD(\Phi,x_i/y_i)$
0	0.42857436	0.86504214	0.39523486	2.1886787779571020	8.99807125	0.92913676	2.49713396	0.07774339	-0.26072569
1	0.44419936	0.85876125	0.40870234	2.1011899433332600	9.04418259	0.94689513	2.50993073	0.07922928	-0.22994397
2	0.45982436	0.85227071	0.42207005	2.0192636438835930	9.08423722	0.96314027	2.52104664	0.08058855	-0.19870098
3	0.47544936	0.84557210	0.43533472	1.9423493314829 70	9.11819601	0.97785502	2.53047084	0.08181977	-0.16697066
4	0.49107436	0.83866706	0.44849311	1.8699664398470 90	9.14602581	0.99102546	2.53819414	0.08292178	-0.13472565
5	0.50669936	0.83155726	0.46154200	1.8016935937725 60	9.16769944	1.00264091	2.54420898	0.08389368	-0.10193720
6	0.52232436	0.82424446	0.47447821	1.7371597543964 20	9.18319574	1.01269395	2.54850949	0.08473484	-0.06857502
7	0.53794936	0.81673042	0.48729859	1.6760369072784 40	9.19249958	1.02118045	2.55109148	0.08544493	-0.03460718
8	0.55357436	0.80901699	0.50000000	1.6180339887499 00	9.19560187	1.02809954	2.55195243	0.08602387	0.00000000
9	0.56919936	0.80110606	0.51257934	1.5628918125475 50	9.19249958	1.03345365	2.55109148	0.08647186	0.03528215
10	0.58482436	0.79299954	0.52503355	1.5103788096549 00	9.18319574	1.03724846	2.54850949	0.08678938	0.07127694
11	0.60044936	0.78469943	0.53735957	1.4602874332178 70	9.16769944	1.03949291	2.54420898	0.08697718	0.10802432
12	0.61607436	0.77620774	0.54955440	1.4124311104580 90	9.14602581	1.04019917	2.53819414	0.08703628	0.14556666
13	0.63169936	0.76752655	0.56161507	1.3666416468712 20	9.11819601	1.03938260	2.53047084	0.08696795	0.18394898
14	0.64732436	0.75865798	0.57353863	1.3227670062872 90	9.08423722	1.03706174	2.52104664	0.08677376	0.22321919
15	0.66294936	0.74960420	0.58532217	1.2806694047792 60	9.04418259	1.03325822	2.50993073	0.08645551	0.26342832
16	0.67857436	0.74036741	0.59696281	1.2402236678295 60	8.99807125	1.02799673	2.49713396	0.08601527	0.30463079

Table 11.2
Inscribed Rectangle Results for
Inscribed Rectangles of Different Aspect Ratios

The Rhombic Triacontahedron[11.4,11.5,11.6,11.7,11.8,11.9]

The Rhombic Triacontahedron is a regular faceted geometrical solid with thirty rhombic sides of equal area. To be precise, each of the identical sides is a Phidian Rhombus, whose long diagonal is Φ and whose short diagonal is unity. The Rhombic Triacontahedron is not one of the (five) Platonic solids because, though its sides are of equal area, the angles included by the lateral edges are not equal: They form two distinct values, as do the internal angles of the Phidian rhombus.

Technically, this Phidian Rhombic Triacontahedron is a Catalan Solid.

The Rhombic Triacontahedron, which some workers abbreviate as RT or r.t., is a combination of a dodecahedron and an icosahedron, and indeed all five of the Platonic solids may be inscribed within the rhombic triacontahedron such that all their vertices touch the appropriate rhombic triacontahedron vertices.

The Dodecahedron, the Icosahedron and the Rhombic Triacontahedron are all haunted by Phi, literally at every turn.

Therefore, twelve of the RT vertices touch the surface of the Outer Circumscriptive Sphere and the remaining twenty touch the Inner Circumscriptive Sphere.

Note that the {x,y,z} Vertex Radii, of which 12 have a common value and 20 a different common value are sensitive to the choice of Edge Length, as are the consequential Surface Area and Volume.

Figure 11.4 is a two-dimensional rendition, shaded to depict this three-dimensional mathematical object.

Figure 11.5 is a wire-frame depiction with its thirty-two labeled vertices. The rhombic triacontahedron has sixty edges.

In a third wire-frame representation I have colored three contiguous *surface* sides to clarify the surficial geometry. This is shown as Figure 11.6

Figure 11.6 offers the reader an optical illusion.

Please note that although the solid triacontahedron contains at least three inscribable cubes the colored sides do ***not*** form an example: As always, beware of pareidolia!

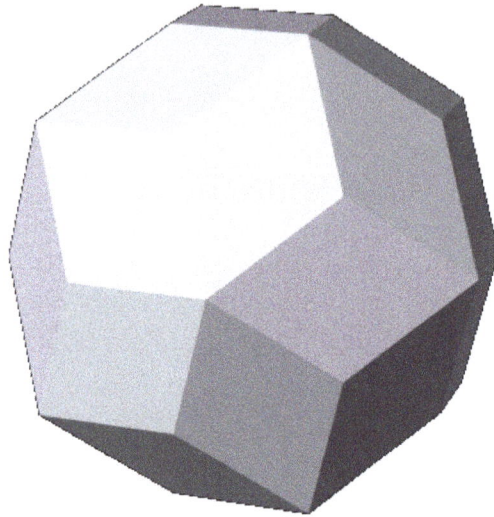

Figure 11.4
The (Phidian) Rhombic Triacontahedron

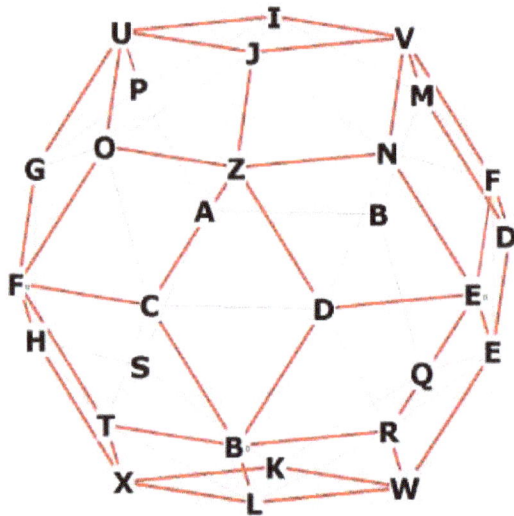

Figure 11.5
The (Phidian) Rhombic Triacontahedron
Wire-Frame Representation with
Labeled Vertices (some occluded)

Dimensions and Metrics

The key datum in the computation of polyhedron metrics is the edge length, at least for those polyhedra like the rhombic triacontahedron, whose edges are equal in length.

Some workers set the Edge Length (s) to unity or something, but I prefer, since we are dealing with Phidian Rhombi, to use the native proportions of that to set the edge length. In context, unity is therefore the Short Diagonal of a Side, and Φ the Long Diagonal. Therefore by reference to Pythagoras:-

$$s = \sqrt{\left(\frac{\Phi}{2}\right)^2 + \left(\frac{1}{2}\right)^2} = 0.951056516295154$$

Equation 11.44

When working with physical values you can of course measure the Edge in meters or something and work with that: The Rhombic Triacontahedron is proportionable.

As implied above, all the metrics of the rhombic triacontahedron will reflect the chosen datum (usually edge) length: So seldom will two workers come up with the same values, especially with regard to Vertex Radii.

Sensible choices of the datum length at or near unity will result in a surface area near to 24 square units and a solid volume of about 10.5 cubic units.

None of this, of course, affects the objective properties of the rhombic triacontahedron as a mathematical object.

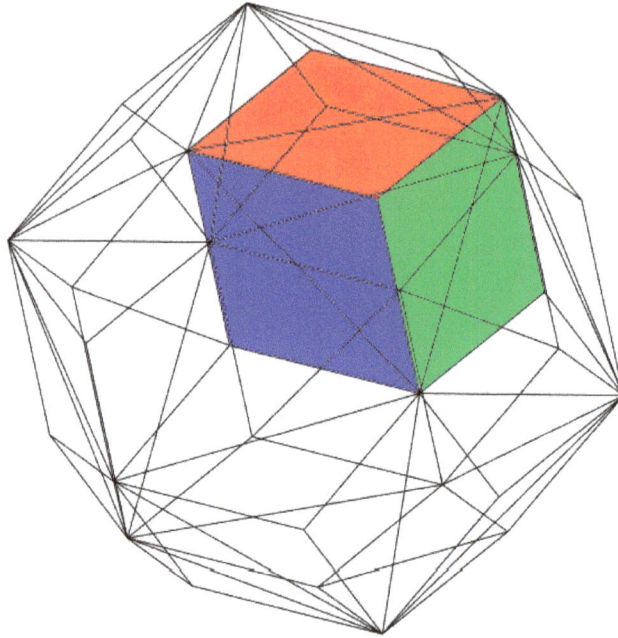

Figure 11.6
The (Phidian) Rhombic Triacontahedron
Wire-Frame Representation with
Three of its Sides Colored

To simplify it is readily ascertained that:-

$$s = \sqrt{\left(\frac{\Phi}{2}\right)^2 + \left(\frac{1}{2}\right)^2} = \Phi\left(\frac{\sqrt{\Phi + 2}}{2\Phi}\right) = \frac{\sqrt{\Phi + 2}}{2} = 0.951056516295154$$

Equation 11.45

The Face Diagonals and the Face Area[11.10,11.11]

Because we have elected that each face of the rhombic triacontahedron is to be considered a golden rhombus it follows that the Major Diagonal, D, is:-

$$D \equiv \Phi$$
Equation 11.46

and that the Minor Diagonal, d, is unity.
Accordingly the Face Area, A_{face}, is:-

$$A_{face} = \frac{dD}{2} = \frac{1 \times \Phi}{2} = \frac{\Phi}{2}$$
Equation 11.47

Please note that if 1 and Φ are dimensionally lengths then Equation 11.47 *implies* an area (L^2). The 1 and the Φ are indeed each lengths (L) in feet or meters or whatever: Vide the Huntley's Cone exercise.

Therefore the Surface Area of the Rhombic Triacontahedron, S, is given by:-

$$S = n. A_{face} = 30 \frac{\Phi}{2} = 15\Phi$$
Equation 11.48

Because the rhombic triacontahedron has n = 30 identical faces.

More generally we may write:-

$$S = 12\sqrt{5}. s^2$$
Equation 11.49

The Volume

The Volume of the Rhombic Triacontahedron, V, is:-

$$V = \frac{5}{2}. \Phi^3$$

Equation 11.50

or in general:-

$$V = 4\sqrt{5 + 2\sqrt{5}}.\,s^3$$
Equation 11.51

If Φ is dimensionally a Length (L) then equation 11.50 is sound.

It is possible to review the identities of the RT volume, for the light they may throw upon further issues, in the following terms:-

$$V = \frac{20}{(3 - \Phi)^{\frac{3}{2}}} \times s^3$$

$$= \frac{20}{(3 - \Phi)^{\frac{3}{2}}} \times \sqrt{\left(\frac{\Phi}{2}\right)^2 + \left(\frac{1}{2}\right)^2}^{\,3}$$

$$= \frac{5}{2} \times \left(\frac{\Phi^2 + 1}{3 - \Phi}\right)^{\frac{3}{2}}$$

$$= \frac{5}{2} \times \Phi^3$$

$$= \frac{5}{2} \Phi(\Phi + 1)$$
Equation 11.52

Alternative Renditions of Surface Area and Volume[11.12]

Wikipedia shows that with the use of native s as given by Equation 11.45 we may write:-

$$S_{wiki} = 12\sqrt{5}.\,s^2 = 12\sqrt{5}.\left[\left(\frac{\Phi}{2}\right)^2 + \left(\frac{1}{2}\right)^2\right] = 3\sqrt{5}(\Phi^2 + 1)$$
Equation 11.53

or as a further alternative but using the Minor Ratio ϕ:-

$$S_{wiki} = 3\sqrt{5}(\phi + 3)$$
Equation 11.54

Meanwhile, if advantageous, the Volume of the Rhombic Triacontahedron may be re-expressed as its general equation:-

$$V_{wiki} = 4\sqrt{5 + 2\sqrt{5}} \times s^3$$

Equation 11.55

Bounding and Approximating Spheres

As we shall later study in some detail, there are two Vertex Radial Distances, $R_{InSphere}$ and $R_{OutSphere}$, that attach to the Rhombic Triacontahedron. The shorter radius, $R_{InSphere}$, characterises twenty radii, and $R_{OutSphere}$ the other twelve.

Thirdly, there is a geometrical Mid-Radius, $R_{midradius}$ of intermediate length.

All these radii are of dimension Length (L).

The three radii are computed using:-

$$R_{InSphere} = \frac{\Phi^2}{\sqrt{\Phi^2 + 1}} \times s = s\sqrt{1 + \frac{2}{\sqrt{5}}}$$

Equation 11.56

$$R_{midradius} = s\sqrt{1 + \frac{1}{\sqrt{5}}} = s\frac{(5 + \sqrt{5})}{5}$$

Equation 11.57

Now $R_{midradius}$ is not necessarily an arithmetic mean, but if we treat it as one we may access an outer bounding radius, $R_{OutSphere}$, in the following terms:-

$$R_{OutSphere} = 2\left(R_{midradius} - \frac{R_{InSphere}}{2}\right)$$

Equation 11.58

So the Volume of the respective inscribed and circumscribed spheres, and that of the Nominal Mean Sphere, $V_{midradius}$, are respectively:-

$$V_{InSphere} = \frac{4}{3}\pi R_{InSphere}{}^3$$

Equation 11.59a

$$V_{midradius} = \frac{4}{3}\pi R_{midradius}{}^3$$

Equation 11.59b

$$V_{OutSphere} = \frac{4}{3}\pi R_{OutSphere}{}^3$$

Equation 11.59c

whist the corresponding Surface Areas of the bounding and central spheres are given by:-

$$S_{InSphere} = 4\pi R_{InSphere}{}^2$$
Equation 11.60a
$$S_{midradius} = 4\pi R_{midradius}{}^2$$
Equation 11.60b
$$S_{OutSphere} = 4\pi R_{OutSphere}{}^2$$
Equation 11.60c

The numerical values of these metrics are summarised in Table 11.3

The Rhombic Triacontahedron: Vertex Co-ordinate Summations[11.13]

Radius/Edge Ratios

Notwithstanding the meaning of contingent spheres, when we quantify the Central Radius of the Inner Vertices, R_{min}, and the Central Radius of the Outer Vertices, R_{max}, we discover that the pairs of the brace differ. And that this difference results in four different numerical values.
To be explicit:-

$$R_{min} = \sqrt{3} = \sqrt{\phi^2 + \Phi^2}$$
Equation 11.61

Major Ratio of Phidias	Φ	1.618033988749890
Minor Ratio of Phidias	φ	0.618033988749895
Edge Length	s	0.951056516295154
Minimum Vertex Radius	R_{min}	1.732050807568880
Maximum Vertex Radius	R_{max}	1.902113032590310
Ratio R_{min} to s	R_{min}/s	1.821185994620060
Ratio R_{max} to s	R_{max}/s	2.000000000000000
InterVertex Depth	d	0.170062225021430
InSphere Radius	$R_{InSphere}$	1.309016994374950
Midradius	R_{mid}	1.376381920471170
OutSphere Radius	$R_{OutSphere}$	1.443746846567400
Count of Minimum Radius	n_{Rmin}	20
Count of Maximum Radius	n_{Rmax}	12
InSphere Volume	$V_{InSphere}$	9.395598801078510
Triacontahedron Volume	V	10.590169943749500
OutSphere Volume	$V_{OutSphere}$	12.605548719326000
Surface Area	S	24.270509831248400
Surface to Volume Ratio	S/V	2.291796067500630

Table 11.3
Dimensions and Metrics of the
Rhombic Triacontahedron
For Edge Length the same as the Golden Rhombus

and:-

$$R_{max} = \sqrt{\Phi^2 + 1}$$
Equation 11.62

	Vertex Serial Number	Perm	Vertex Co-Ordinates			Vertex Radius
			0.0000	0.0000	0.0000	
			x	y	z	
1	1	(0,+1,+Φ)	0.0000	1.0000	1.6180	1.9021
	2	(⁺1,⁺Φ,0)	1.0000	1.6180	0.0000	1.9021
	3	(+Φ,0,1)	1.6180	0.0000	1.0000	1.9021
	4	(0,+1,+Φ)	0.0000	1.0000	1.6180	1.9021
	5	(0,−1,−Φ)	0.0000	-1.0000	-1.6180	1.9021
	6	(−1,−Φ,0)	-1.0000	-1.6180	0.0000	1.9021
	7	(−Φ,0,−1)	-1.6180	0.0000	-1.0000	1.9021
	8	(0,−1,−Φ)	0.0000	-1.0000	-1.6180	1.9021
	9	(0,+1,−Φ)	0.0000	1.0000	-1.6180	1.9021
	10	(+1,−Φ,0)	1.0000	-1.6180	0.0000	1.9021
	11	(−Φ,0,1)	-1.6180	0.0000	1.0000	1.9021
	12	(0,⁺1, ⁻Φ)	0.0000	1.0000	-1.6180	1.9021
2A	13	(1,1,1)	1.0000	1.0000	1.0000	1.7321
	14	(1,1,-1)	1.0000	1.0000	-1.0000	1.7321
	15	(1,-1,1)	1.0000	-1.0000	1.0000	1.7321
	16	(-1,1,-1)	-1.0000	1.0000	-1.0000	1.7321
	17	(-1,-1,-1)	-1.0000	-1.0000	-1.0000	1.7321
	18	(-1,-1,1)	-1.0000	-1.0000	-1.0000	1.7321
	19	(-1,1,-1)	-1.0000	1.0000	-1.0000	1.7321
	20	(1,-1,1)	1.0000	-1.0000	1.0000	1.7321
2B	21	(0,+Φ,+1/Φ)	0.0000	1.6180	0.6180	1.7321
	22	(+Φ,+1/Φ,0)	1.6180	0.6180	0.0000	1.7321
	23	(+1/Φ,0,+Φ)	0.6180	0.0000	1.6180	1.7321
	24	(+1/Φ,+Φ,0)	0.6180	1.6180	0.0000	1.7321
	25	(+Φ,0,+1/Φ)	1.6180	0.0000	0.6180	1.7321
	26	(0,+1/Φ,+Φ)	0.0000	0.6180	1.6180	1.7321
	27	(0,−Φ,−1/Φ)	0.0000	-1.6180	-0.6180	1.7321
	28	(−Φ,−1/Φ,0)	-1.6180	-0.6180	0.0000	1.7321
	29	(−1/Φ,0,−Φ)	-0.6180	0.0000	-1.6180	1.7321
	30	(⁻1/Φ, ⁻Φ,0)	-0.6180	-1.6180	0.0000	1.7321
	31	(−Φ,0,−1/Φ)	-1.6180	0.0000	-0.6180	1.7321
	32	(0,−1/Φ,−Φ)	0.0000	-0.6180	-1.6180	1.7321
Sum			-0.6180	0.3820	-4.2361	57.4664
Mean			-0.0193	0.0119	-0.1324	1.7958
Pop. Standard Dev.			0.9975	1.0450	1.1093	0.0836

		$\Sigma=$	$-\phi$	$+\phi^2$	$-1-2\Phi$	
Count	-1		5	6	6	
	0		10	7	7	
	1		6	8	5	
	$-\phi$		2	2	2	
	$+\phi$		2	2	2	
	$-\Phi$		4	4	6	
	$\lvert\Phi$		3	3	4	
Count Totals			32	32	32	

Vertex Serial Numbers and Co-Ordinate Permutations are in No Particular Order

Table 11.4
Rhombic Triacontahedron Vertex Co-Ordinates
For Edge Length the same as the Golden Rhombus

Reference to Table 11.4 confirms that the Sum of z-Vertex Co-Ordinates, Σ_z, is given by:-

$$\begin{aligned} \Sigma_z &= 6 \times (-1) + 7 \times 0 + 5 \times 1 + 2 \times -\phi + 2 \times \phi + 6 \times -\Phi + 4 \times \Phi \\ &= -6 + 5 + 2 \times -\Phi \\ &= -1 + 2 \times -\Phi \\ &= -1 - 2\Phi \end{aligned}$$

Equation 11.63

Because there are twelve long Vertex Radii and twenty short Vertex Radii further reference to Table 11.4 confirms that the Sum of All Vertex Radii, Σ_{VR}, is:-

$$\Sigma_{VR} = 12\sqrt{\Phi^2 + 1} + 20\sqrt{3}$$

Equation 11.64

which may be expanded as:-

$$\Sigma_{VR} = 4\left(3\sqrt{\phi^2 + 2\phi + 2} + 5\sqrt{2\phi^2 + 2\phi + 1}\right)$$

Equation 11.65

The Ratio of the Short Vertex Radius to the Side Length is given by:-

$$\frac{R_{min}}{s} = \frac{\sqrt{3}}{\sqrt{\left(\frac{\Phi}{2}\right)^2 + \left(\frac{1}{2}\right)^2}} = \frac{\sqrt{3}}{\sqrt{\frac{\Phi^2}{4} + \frac{1}{4}}}$$

Equation 11.66

whose square is:-

$$\left(\frac{R_{min}}{s}\right)^2 = \frac{3}{\frac{\Phi^2}{4} + \frac{1}{4}} = \frac{3}{\left(\frac{1}{4}\right)(\Phi^2 + 1)} = \frac{12}{\Phi^2 + 1} = \frac{12}{\Phi + 2}$$

Equation 11.67

and so the ratio R_{min}/s reduces to:-

$$\frac{R_{min}}{s} = \sqrt{\frac{12}{\Phi + 2}}$$

Equation 11.68

Meanwhile the Ratio of the Long Vertex Radius to the Side Length is given by:-

$$\frac{R_{max}}{s} = \frac{\sqrt{\Phi^2 + 1}}{\sqrt{\left(\frac{\Phi}{2}\right)^2 + \left(\frac{1}{2}\right)^2}} = \frac{\sqrt{\Phi^2 + 1}}{\sqrt{\frac{\Phi^2}{4} + \frac{1}{4}}} = 2$$

Equation 11.69

whose square is:-

$$\left(\frac{R_{max}}{s}\right)^2 = \frac{\Phi^2 + 1}{\frac{\Phi^2}{4} + \frac{1}{4}} = \frac{\Phi^2 + 1}{\left(\frac{1}{4}\right)(\Phi^2 + 1)} = 4\frac{\Phi^2 + 1}{(\Phi^2 + 1)} = 4$$

Equation 11.70

CHAPTER TWELVE
PHI IN THE ANIMAL KINGDOM

The Ratio of Phidias, Φ, (The Golden Ratio); the Golden Rectangle; and the Golden Spiral appear most suggestively throughout the Animal Kingdom.

I say suggestively because the human mind is what it is and tries to see pattern and meaning everywhere. But God and his Creatures carry on regardless for a number is a number, a literary creation of the human mind. For sure, we think that Pi and Phi have something to do with the stability of organic growth, but stability and growth are things regardless, and they will persist long after you and I are dead, and men have joined the Ammonites, those named in the leaves of The Bible[12.1] and those very different Ammonites in the layers of the Blue Lias.

The Ammonites of the Blue Lias are common in the rocks of the quaint and atmospheric little port of Whitby on England's cold North-East Coast. In 657AD St Hilda founded a great abbey on the south cliff-crest overlooking the town. The abbey is now in ruins, slighted long ago by Henry the Eighth. Molluskan ammonites take the form of curled silica squids, nautilus-like, and not unlike coiled snakes. Perhaps inevitably the local people averred that they were indeed such, miraculously petrified by Hilda. You take my point?

Gryphaea arcuata

Gryphaea arcuata is a prehistoric oyster who lived on muddy sea floors in the Lower Jurassic some 195 million years ago. The Blue Lias is a succession of sedimentary rocks containing plentiful *Gryphaea*, and is an intercalation of fine muds and sands indurated to a soft, shaly rock that breaks readily on stormy sea coasts such as those of Yorkshire and Dorset in England.

The *Gryphaea* shells, their exoskeleton, have been silicified as fossils, and may easily be found intact on the foreshore, or of course chiseled from the living rock.

My Late Father found a splendid specimen, about five centimeters in length, when he investigated the geology of the foreshore near Whitby[12.2,12.3,12.4].

Before we take a look at *Gryphaea* we will re-visit our Golden Rectangle and its contained version of the Golden Spiral because we may be able to use them as convenient templates in our continuing studies.

Figure 12.0 is our previous Golden Spiral in a Golden Rectangle image which I have edited to remove superfluous squares and other confusing detail:-

Figure 12.1 is an excellent photograph of the sinistral surface profile of an excellent *Gryphaea arcuata* specimen.

Note the concentrated, fine growth rings. If these are annual rings it would not be inconsistent with the age of the specimen: Oysters are capable of surviving for many centuries.

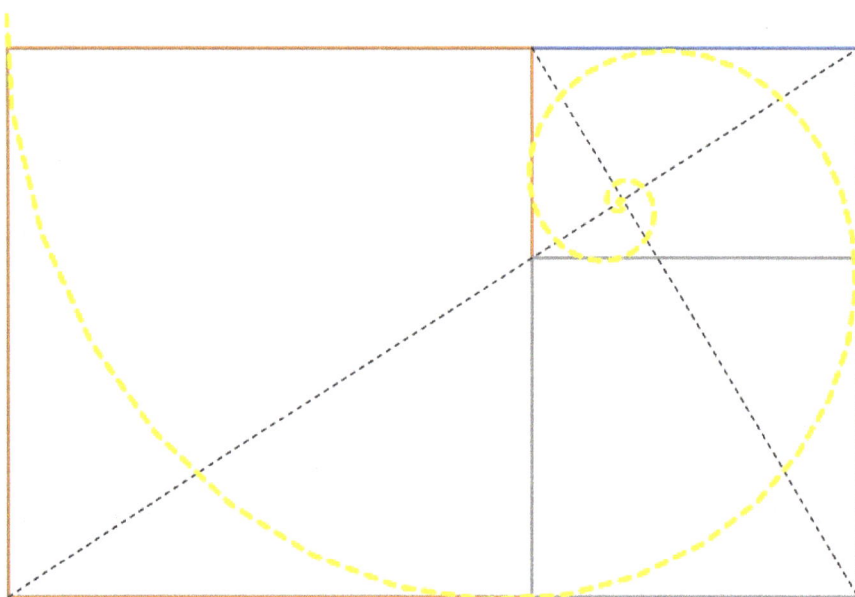

Figure 12.0
Golden Rectangle and Contained Golden Spiral
Image Comparisons Template

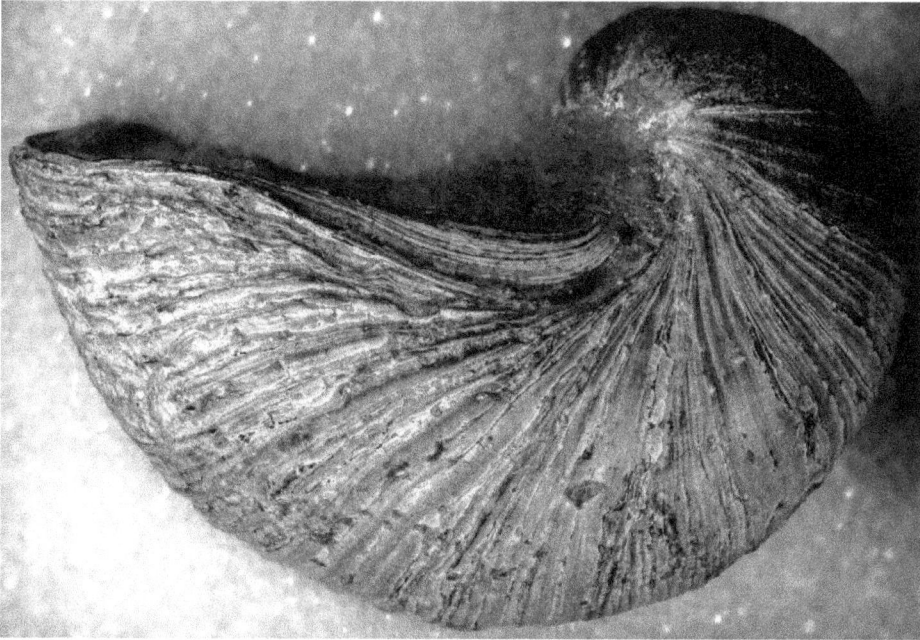

Figure 12.1
Gryphaea arcuata
Left (Sinistral) Outer Profile
Length: 6.4 centimeters

When the template is superposed upon the photograph *without distortion of the aspect* the following striking and very beautiful conformation is apparent:-

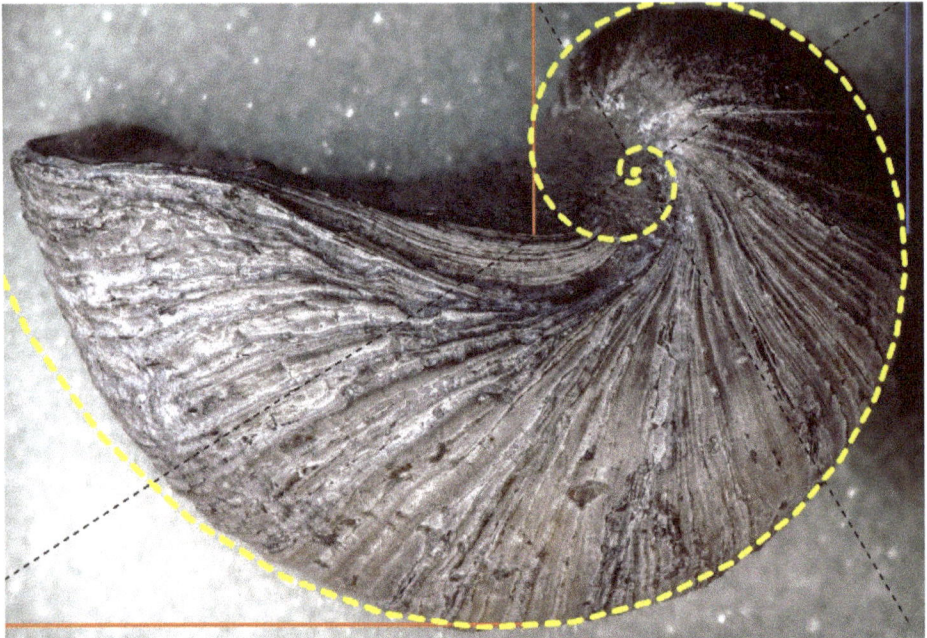

Figure 12.2
The Apparent Golden Spiral Growth Habit of *Gryphaea arcuata*

As we have shown in the description of the Golden Rectangle the equation of this *particular* spiral is:-

$$r_i = 1.94649798 \times e^{\left[-0.1097+\pi-\tan^{-1}\left(\frac{1-y_U}{x_U}\right)+2\pi v\frac{i}{n}\right]}$$
Equation 12.1

where x_U and y_U are respective Umbilical Co-Ordinates and v is the Number of Volutes, which would need to be assessed by a competent palaeontological technician, but is probably about four in this example.

Having said this, the geometry is not outrageously unlike that of that of this single-volute equation:-

$$r_i = x_U.e^{-\phi.\left(2\pi\frac{i}{246}+\frac{\pi}{2}\times\frac{48}{32}\right)}$$
Equation 12.2

which obviously simplifies to:-

$$r_i = x_U \cdot e^{-\phi.\pi\left(\frac{i}{123}+\frac{3}{4}\right)}$$
Equation 12.3

Equation 12.3 describes a curve sometimes called a scaphoid because of its supposedly boat-like profile.

This shows the critical importance of empirical scientific interpretations to the understanding of reality, and the relation of that reality (if any) to mathematical constructs.

Of course, we could digitise the profile and fit a logarithmic equation programmatically, or apply other statistical analyses, but that would not allow us to evade the essential subjectivity of the science.

Now I am very proud of Equation 12.2. I am sure you like it too. I lavished days of fraught labor getting it just *right*, just like other peoples' diagrams of the spiral that intercepted sub-squares' corners in the Golden Rectangle.

And yet and yet ...

The term ψ in the expression:-

$$\psi = \left(2\pi\frac{i}{246} + \frac{\pi}{2} \times \frac{48}{32}\right) = \pi\left(\frac{i}{123} + \frac{3}{4}\right)$$
Equation 12.4

is merely a Rotation. I am sure that if you saw Jim Warren driving a bus, having put on weight since you saw him last year lying on a sunlounger in Ibiza, you would instantly say to yourself "This is the same Jim Warren I saw last year with his feet to the South in Spain".

Dilations, Translations and (rigid) Rotations ("Affine Transformations") are mathematical figments that do not affect the intrinsic properties of the object.

The equation:-

$$r_i = e^{-\phi}$$
Equation 12.5

describes a curve that is very similar to or identical with the Golden Spiral of Equations 12.2 and 12.3 The only difference is the Rotation of ψ and the Dilation of a.

And yet Equation 12.5 describes a circle, and implies Pi. Something critical is missing. That critical, that Necessary and Sufficient added ingredient, is Growth.

Man invented the Ratio of Phidias, and a man called Viète invented the modern algebra for us to play with it, but no man invented *Gryphaea*.

The universe that we inhabit is an emanation of the thought of a sentient Agent which cohabits our space. Our universe may even be an afterthought or a mere by-product like the footnotes of a annalist or a dream in a tale told by a Father. *Gryphaea* is regardless. He was conceived when the morning stars sang together and entered into the springs of the sea. For sure he exists as a silicious cast in a rock or on a beach, and we are operationally certain that he was once a living oyster. Or rather his fossil form reflects the contours of his exoskeleton that rotted long ago to leave a void filled eons later by a mineral that preserved his Form but not his Substance somewhat as the pharaohs of Egypt were conserved. But the mathematics is the work of novelists, a literary projection of our psychoneurological processes and projections by which we endeavor to make sense of our environment and extend our own survival. It is language, and most that is said is either trivial or plain wrong.

Here we once again come up hard against the Higher Pareidolia. "For my thoughts are not your thoughts, neither are your ways my ways, saith the LORD."[12.5]

Ammonites at The American Mathematical Society[12.6]

The logarithmic spiral conformation of Mesozoic ammonites is discussed in an interesting way on the AMS website. There is no implication by either the AMS or by me that the spiral geometry of these ammonites has anything to do with Phidian mathematics.

An ammonite is an extinct pre-historic marine animal who had a curly shell, in a two-dimensional (x-y)-plane somewhat like a snail rather than the helical cones of modern whelks. The animal itself is assumed to have been a squid-like creature of cephalopodal type, similar to the modern Chambered or Pearly Nautilus (*Nautilus pompilius*). The shell was multi-partitioned, and the actual body chamber had a sub-elliptical conformation in the (x-z)-plane. The pearly nautilus is an animal of a highly-evolved type, fit to be mentioned in the same breath as whales and men.

Figure 12.3
The Pearly Nautilus (*Nautilus pompilius*)
Photo: Lee R Berger[12.7]

The AMS treatment is three-dimensional and an elliptical cross-section in the (x-z)-plane is assumed.

If F(x,y,z) is a Point on the Surface of the Preserved Exoskeleton than the relevant parametric equation specified is:-

```
plot3d([exp(v*t)*(h+a*cos(s))*cos(t),-
exp(v*t)*(h+a*cos(s))*sin(t),
exp(v*t)*(k+b*sin(s))],s=0..2*Pi,t=-10..-
1,grid=[30,50],scaling=constrained)
```

in the Maple idiom.

I have tried to transliterate the Maple as:-

$$x_{s,t} = e^{vt.(h+a.\cos(s)\cos(t))}$$
Equation 12.6a

$$y_{s,t} = -e^{vt.(h+a.\cos(s)\sin(t))}$$
Equation 12.6b

$$z_{s,t} = e^{vt.(k+b.\sin(s))}$$
Equation 12.6c

For the case of *Astroceras obtusum*, a fossil ammonite common in the Lower Jurassic of Dorset the relevant parameters are:-

Astroceras.
v=.12, c=1, a=1.25, b=1.25, h=3.5, k=0,t=-40..-1

Figures 12.4 and 12.5 are respectively the photographs and Maple wire models of *obtusum* provided upon the relevant AMS webpage.

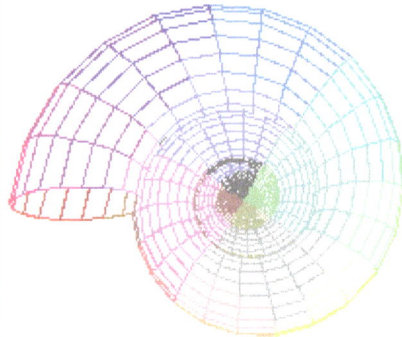

Figures 12.4 and 12.5
Photograph and Maple Model
of *Astroceras obtusum* from
the Lower Jurassic of Dorset
Photo, Model: AMS contributors

The modeler admits that he adjusted the seven parameters to fit the model "by eye" (clearly with a much higher standard of craftsmanship than I attained with the *Gryphaea* and its Golden Spiral). But a fitted model, even if it is fitted using "objective" statistical software, is only a model and propagates the perceptions and the prejudices of its promulgator.

Ammonites in Old Literature

Ammonites and their seductive forms appear frequently in paleontological research papers of the past, dating back to the Forties of the Twentieth-Century and even well before.

Many of these studies were very meticulous from the point of view of both scholarship and laboratory measurements, but were sadly lacking in mathematical rigor.

Indeed, the late Dr Ethel Currie presented numerous linear log-normal plots of ammonite whorl (volute) growth metrics, and then flatly denied that ammonites exhibited logarithmic spiral forms! (In particular, she concludes that "*The umbilical spiral is never a true logarithmic spiral*", true in the letter, but somewhat misleading).[12.8,12.9]

Other interesting contributions include those of Bucker, Landman, Klofak and Guex who in a 1996 book Chapter 12 entitled "Mode and Rate of Growth in Ammonoids" advanced several exponential rules of the pattern:-

$$f(x) = e^{a+b.x}$$
Equation 12.7

to describe the relations of various metrics of the ammonite anatomy. For example, my deconstruction of Figure Three that relates Septum Interdistance in mm (y) to Number of Septa (x) in a juvenile specimen of *Parafrechites meeki* (Middle Triassic, Nevada, HB 23745) takes the form:-

$$y = e^{-2.32451008+0.04448655x}$$
Equation 12.8

But Bucher Et Al give many exponential-logarithmic spacial relations explicitly, such as:-

$$r = ae^{k\vartheta}$$

Equation 12.9

where r is the Planimetric Radius of the specimen, where r$_{\theta\text{-}n\pi}$ is the Radius at the nth. Stage of Ontogeny; and θ is the Ontogenic Lapse Angle.

Or:-

$$W = e^{2k\pi}$$

Equation 12.10

where W is the Rate of Whorl (i.e. Volute) Expansion.

Many readers must think that I have spent too much time discussing ammonites irrelevant to the themata of Phidian mathematics, but the point that I insist upon is that the Golden Spiral is a very specialised logarithmic spiral, and that the many animals who incorporate logarithmic growth patterns into their anatomies have no scientific or mystical reason to favor the Golden Spiral in particular.

Ethics

You may perhaps agree that we have little scruple to consult the convenience or serenity of dead oysters or dead squids. Living, feeling creatures are a different matter entirely. Even students.

There exists an Indian-German study in which 479 high school students of both sexes where required to be measured for interdistances that included height-to-scalp and feet-nipple-navel ratios.

During this "fascinating exposition" (the authors' words, not mine) the assertion of Leonardo da Vinci that the ratio of height of a man to his sole-to-navel distance was Φ was noted, and confirmed precisely for the 207 German students. However, the Indians fell short (so to say) at a mere 1.615

After ritual genuflections in the direction of dead geniuses as diverse as Leonardo Pisano, Le Corbusier, and Fibonacci; The Parthenon; and the rehearsal of the old error about the Great Pyramid of Giza, plus statistical tests of significance and other tokens of respectable credibility the authors conclude that nipples have nothing to do with Phidian theory.

One may of course legitimately question the independent volition of children, or even "young men" in a statistics institute, and I am sure this 1973 study would not be approved today, and am categorically certain that it would not be published.

But having said that, as recently as 2021 a Jordanian study[12.10] used 380 undergraduates of both sexes to attempt to relate various averaged anatomical lengths to a rag-bag of independent variables including sex, academic attainments, BMI, blood group, and you name it to the ideal of "Vitruvian Man". Sophisticated statistical analyses were applied and of course it was demonstrated that everyone fell short, especially if they were female, obese or stupid.

Different, but related issues of consent and benefit, are raised by a paper submitted in a much more rigorous and respectful spirit by a team of British orthopedic paediatric surgeons.

We shall discuss this last submission in some detail.

Distal Physes of the Human Radius and Ulna[12.11]

The distal physes are cartilaginous growth pads at the wrist end of the human arm. They extend the lower arm bones, the radius and the ulna. The ulna is the more slender of the two bones.

Mamarelis Et Al formed the opinion that the total normal width of the radial and ulnar growth pad combined was in the Major Phidian Ratio to the normal width of the radial growth pad alone. They accordingly consulted the hospital records of 268 patients of ages 3 to 16 in case they threw light upon this question. Mamarelis and his colleagues found that the mean of the ratio in evidence was 1.619684. Meanwhile, the relevant standard deviation was computed to be 0.0179473. The Marmarelis Team did not specify whether they applied σ, the Population Standard Deviation, or s, the Sample Standard Deviation: The two figures differ by 0.373%..

To be specific, the Team's thesis was:-

$$\Phi \approx \frac{r_r + r_u}{r_r} = \frac{r_r}{r_u}$$
Equation 12.11

where r_r is the geometrical Radius of the Radius Bone growth pad and r_u is the Geometrical Radius of the Ulna bone growth pad.

Because Φ is 1.61803398874989 I have calculated that the Percentage Specific Defect, PSD(Φ,Φ_{MarmarelisEtAl}) is -0.101976303438762, so that the discrepancy between Φ and the ratio of the two pads is around one part in one thousand. You are capable of assessing this finding, and I leave comment your prerogative.

As I moved further into the Team's disquisition I was increasingly disquieted by the numerous radiographs of infant wrists. Irradiation of living tissue is never beneficial and I wondered if the pictures were justified, especially for the happy but ultimately trifling sake of Phi.

But Mamarelis and his colleagues assure us that they only consulted *existing records* of young patients who had presented for interventions with wrist fractures between January 2010 and January 2017. The Team further assured us that these historical radiographs, though they post-dated restorations, were only taken for purposes of clinical confirmation, without a research end in view.

At the end of their report, Mamarelis and his colleagues make the following espousals, which I quote verbatim:-

Disclosures

Human subjects: *Consent was obtained or waived by all participants in this study.* **Animal subjects:** *All authors have confirmed that this study did not involve animal subjects or tissue.* **Conflicts of Interest:** *In compliance with the ICMJE uniform disclosure form, all authors declare the following:* **Payment/services info:** *All authors have declared that no financial support was received from any organization for the submitted work.* **Financial relationships:** *All authors have declared that they have no financial relationships at present or within the previous three years with any organizations that might have an interest in the submitted work.* **Other relationships:** *All authors have declared that there are no other relationships or activities that could appear to have influenced the submitted work.*

Though I naturally doubt the ability of a three-year-old or even a nine-year-old to give informed consent to the divulgence of medical records, and do not share their optimism concerning the clinical utility of their findings, I nevertheless consider the Mamarelis Team study to be a model for British science.

Of course the Disclosures declaration is naive and as legally-watertight as a colander, but it is a step in the right direction.

All data were of course anonymised.

Figure 12.6 reproduces the radiograph "Figure 9" of the Mamarelis Et Al paper and I include it because it provides definitional lineaments for the Mamerelis mathematics.

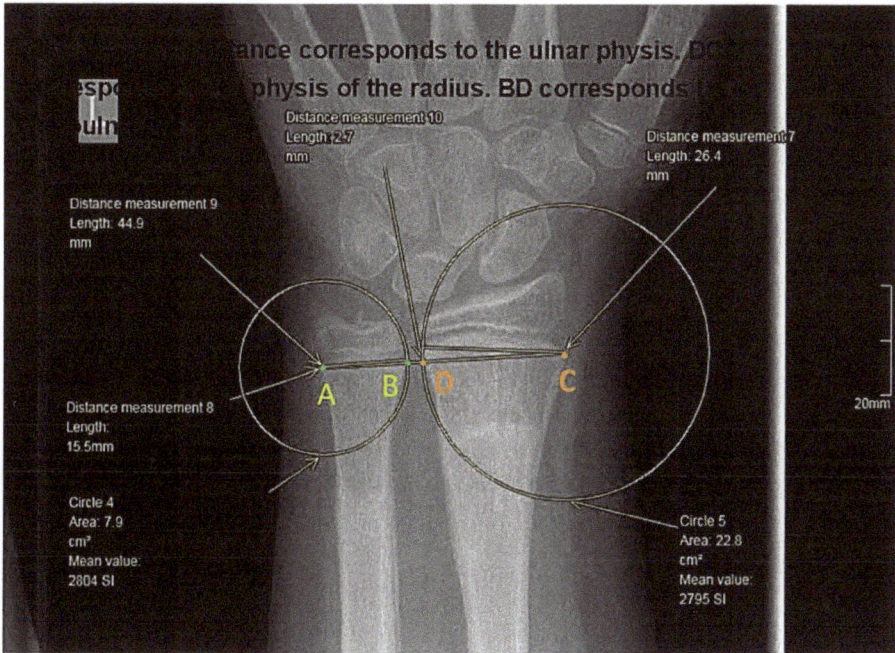

Figure 12.6
Definitional Radiograph of a Human Wrist
Photo: Mamarelis, Karam, Sohail and Key
"AB distance corresponds to the ulnar physis.
DC corresponds to the physis of the radius.
BD corresponds to the radioulnar joint."

The accompanying equation given by the Team is:-

$$\frac{CD + BD/2}{AB + BD/2} = 1.619684 \approx \Phi$$
Equation 12.12

Usefully, the Mamarelis Team also provide a summary of their results in the form of "Table 1", a Grouped Frequency Distribution (GFD) of the Mean Ratios by Age Year. I extended this to show PSD(Φ,$\mu_{Mamarelis}$) for each Age and also some Simple and GFD Means and Standard Deviations.

I present these findings in Table 12.1

The Mean for a Grouped Frequency Distribution is:-

$$\mu_{GFD} = \frac{\sum fx}{\sum f}$$
Equation 12.13

and the Standard Deviation for a Grouped Frequency Distribution is:-

$$\sigma_{GFD} = \sqrt{\frac{\sum fx^2}{\sum f} - \left(\frac{\sum fx}{\sum f}\right)^2}$$
Equation 12.14

where x is the Mid-Point of the (equal) Interval (in this case Age), and f is the Number of Patients in the Interval.

I have also provided Medians where meaningful.

As you can see, there are minor discrepancies between the Simple and GFD statistics due to the loss of information (data blurring) implied by grouped frequency treatments.

Figure 12.7 is a histogram of the Patients' Age Distribution. As you can see it is multimodal but somewhat skewed towards the seven- or eight- year-olds who inevitably run around a lot with minimal supervision.

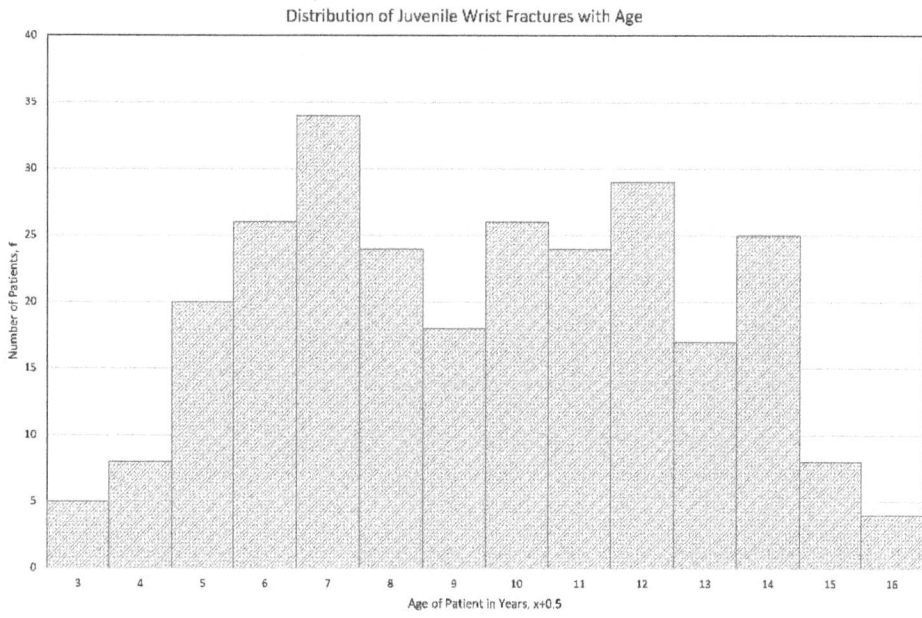

Distribution of Juvenile Wrist Fractures with Age

Figure 12.7
Patient Age Distribution

Φ   1.618033989

Age	Number of Patients	Mean Measured Ratio	PSD $(\Phi, \mu_{mamarelis})$	Age Mid-Point (x)	fx	fx²	
3	5	1.618912731	-0.054309258	2.5	12.5	31.25	
4	8	1.615254249	0.171797365	3.5	28	98	
5	20	1.621134699	-0.191634432	4.5	90	405	
6	26	1.617770116	0.016308233	5.5	143	786.5	
7	34	1.621147302	-0.192413341	6.5	221	1436.5	
8	24	1.620943379	-0.179810206	7.5	180	1350	
9	18	1.611171123	0.42148429	8.5	153	1300.5	
10	26	1.623010678	-0.307576311	9.5	247	2346.5	
11	24	1.629361125	-0.70005552	10.5	252	2646	
12	29	1.621080288	-0.188271648	11.5	333.5	3835.25	
13	17	1.616227367	0.111655365	12.5	212.5	2656.25	
14	25	1.615250012	0.172059226	13.5	337.5	4556.25	
15	8	1.610687303	0.45405015	14.5	116	1682	
16	4	1.626401356	-0.517131736	15.5	62	961	
Total	268			126	2388	24091	
Simple μ	9.5	19.14286	1.619167981	-0.070084549			
Simple σ	4.031129	9.171829	0.005090528	0.31461194			
Median	9.5	22	1.619928055	-0.117059732	8.91044776			GFD μ
					3.23970859			GFD σ

Table 12.1
Mamarelis Et Al GFD of Juvenile Wrist Physis Ratios

CHAPTER THIRTEEN
PHI IN THE VEGETABLE KINGDOM

The Golden Angle and the Fibonacci Series are both key to the growth and anatomy of plants, and therefore so is the Ratio of Phidias.

We know that the (Major) Ratio of Phidias, Φ, has the equation $(1+5^{\frac{1}{2}})/2 \approx 1.618033988749890$, and that the Golden Angle is given by:-

$$\gamma = 2\pi f = 2\pi \left(1 - \frac{1}{\Phi}\right) = 2\pi(1 - \phi) = \frac{2\pi}{\Phi^2} = \pi\left(3 - \sqrt{5}\right)$$
Equation 13.1

and that it therefor follows that the Golden Angle, γ, is itself transcendental, has no rational multiplier to fit the circuit, but has a numerical value (in radians) of about 2.39996322972865

We also know the Laws of God are not the laws of men, and that His prescriptions are a Guide not a constraint.

If a component of a flower, a cactus or a tree is generated radially and one-by-one or in clusters, simultaneously with the upward growth of the plant, then orderly, but not exactly invariant patterns arise.

We can see this rule exemplified by the numbers of petals on a flower, which is usually a Fibonacci Number, typically five, as demonstrated in Table 13.1

The bold red "NF" is a petal count that is *not* a Fibonacci Number. But on the whole petal counts appear to be Fibonacci Numbers.

Let Miss Sasha Langholz[13.1] counsel us:-

"2/3 or 66% of Asteraceae flowers had a Fibonacci number of petals. 12/13 or 92% of the Myrtaceae flowers had a Fibonacci number of petals. 8/8 or 100% of the cones of the Pinaceae family had Fibonacci numbers of spirals....

....The Asteraceae family had the least Fibonacci numbers. In books, the aster family is recognized as the family with the most Fibonacci numbers.

For example, Ian Stewart in Nature's Numbers says "In nearly all flowers, the number of petals is one of the numbers that occur in the sequence 1, 1, 3, 5, 8, 13, 21, 34, 55, 89. For instance lilies have 3 petals, buttercups have 5, delphiniums have 8, marigolds have 13, asters have 21, and most daisies (asters) have 34, 55, or 89."

This is not true. Aster petals show some Fibonacci numbers but aren't consistent. 93% of Myrtaceae flower petals are Fibonacci, but most had 3 or 5 petals. This does not show conclusively that Fibonacci numbers occur in myrtles, it could just be a plant characteristic. Fibonacci numbers appear in plants, more often in the Pinaceae family and less in Asteraceae and Myrtaceae."

I do not know how old Miss Langholz is, but she acknowledges her uncle for introducing her to Fibonacci and her "mom" for finding her "reliable research" and for "driving me around". Even if the lady proves to be fifty-seven her research is as excellent as her science.

Phyllotaxis

Phyllotaxis is the arrangement of leaves during the growth of a plant.

A closely related, but distinct, aspect is parastichy, defined by Wikipedia [13.2] as:-

Parastichy, in phyllotaxy, is the spiral pattern of particular plant organs on some plants, such as areoles on cacti stems, florets in sunflower heads and scales in pine cones.[1] These spirals involve the insertion of a single primordium.

The European Sneezewort *Achillea ptarmica* exemplifies several Phi-related patterns in its growth and phyllotaxis. Sneezewort has between 7 and 15 petals in each flower.

Figure 13.1
The Sneezewort
Achillea ptarmica

Number Of Petals	Flower	Family Name	Fibonacci Number
1 White Calla Lily			1
2 Euphorbia			2
3 Lily, Iris		Iridaceae	3
4 Fuchsia			NF
5 Primrose, Buttercup, Larkspur, Wild Rose, Columbine, Pink, Larkspur, Hibiscus		Rosaceae	5
6 Crocus, Narcissus, Amaryllis			NF
8 Delphiniums, Bloodroot			8
13 Ragwort, Corn Marigold, Cineraria		Asteraceae	13
21 Aster, Black-Eyed Susan, Chicory			21
34 Plantain, Pyrethrum, Daisy			34
55 Michaelmas Daisies, Asteraceae family			55
89 Michaelmas Daisies, Asteraceae family			89

Table 13.1
The Petal Count of Some Flower Families

The sneezewort exhibits, at stages of its stems' bifurcations, generations of leaves whose sub-population is a Fibonacci Number. The stagewise proliferation pattern is illustrated below (after Huntley)[13.3]:-

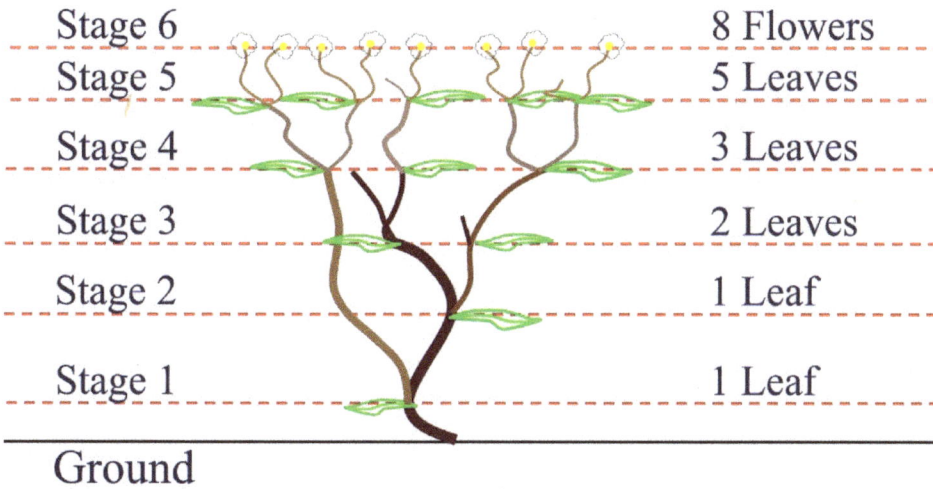

Stage 6	8 Flowers
Stage 5	5 Leaves
Stage 4	3 Leaves
Stage 3	2 Leaves
Stage 2	1 Leaf
Stage 1	1 Leaf

Ground

Figure 13.2
The Stagewise Production of Leaves
in the
European Sneezewort *Achillea ptarmica*

Like many diagrams and description in natural history, Figure 13.2 is very informative as long as it is not taken too literally. On the whole, Fibonacci is the principle of the thing, and the weed culminates in a Fibonacci spray of flowers, as do many other vegetable species.

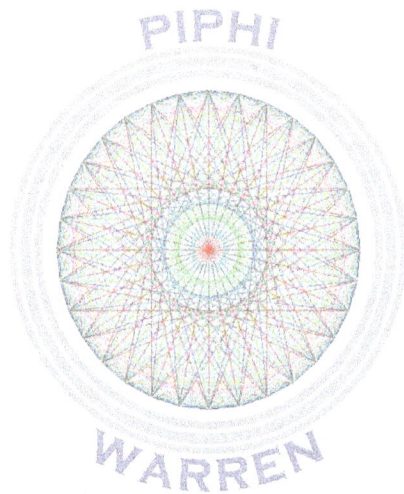

CHAPTER FOURTEEN
PHI IN THE MINERAL KINGDOM

Jack sidled up to me with an unaccustomed air of diffidence. He carried in his hands a small book and furtively opened it to a pre-selected page. He disclosed a shocking monochrome photograph.

It depicted what appeared to be a smooth surface: Jack said of ground gold. On the pristine and perfect plain of precious metal sat pat a perfect pentagonal pyramid of crystal metal gold.

Jack showed me the picture nearly ten years before Dan Shechtman discovered quasicrystals on 8 April 1982, a coup for which Dan won the 2011 Nobel Prize in Chemistry.

Jack need not have bothered.

The picture was already the gossip of the Department, but whispered in the hushed and hesitant tones of scandal, as all heresies are.

Of course, most of us suspected fraud or hoax.

"What do you think of this, Jim?" asked Jack.

I stood and smirked inanely.

How could I advise this man? What could a twenty-year-old tell one of the most practiced crystallographers alive about gold crystals?

You see, once again, even after the lapse of more than fifty years, it is not my memory but my powers of relation that fail me.

Let Professor Liden take up the story.

The problem is this[14.1]:-

"Solid state matter always displays substantial short-range and long-range order to various degrees. Short-range order is imposed by the typical local bonding requirements of chemistry. Even in materials such as silica glass, which are normally considered to be completely amorphous, substantial local order is present: Each silicon atom is tetrahedrally surrounded by four oxygen atoms at 1.62 Å, and the typical oxygen–oxygen separation is 2.65 Å. While it is possible to detect one or two more distinct structural traits in silica glass, the material lacks the hallmark of crystallinity: long-range order.

Ever since the work of Abbé Haüy in 1784, where he showed that the periodic repetition of identical parallelepipeds (molécules intégrantes, now known as unit cells) can be used to explain the external shape of crystals, long-range order has been assumed to be inextricably linked to translational periodicity. Hence, the classical definition of a crystal is as follows: A crystal is a substance in which the constituent atoms, molecules, or ions are packed in a regularly ordered, repeating three-dimensional pattern. Implicitly, this means that a crystal is infinite, and given the size of the unit cell (tens to hundreds of ångströms) in comparison to the size of the physical crystal (hundreds of microns), practically, this is not too far off the mark. The vast majority of the unit cells form the bulk of the crystal, and only a very small part form the surface.

Real crystals are, of course, not only finite in size but also contain imperfections, and the borders between crystalline and amorphous are to some extent defined by the measurement method. A sample that appears crystalline to a local probe such as selected area electron diffraction may appear amorphous to powder X-ray diffraction. One of the most striking characteristics of crystals is their space-group symmetry. The 230 space groups were enumerated in the late 19th century independently by Fedorov, Barlow and Schoenflies. Many local symmetry operations that are incompatible with translational symmetry may still be realized by isolated molecular assemblies. Among the rotational symmetries, 2-, 3-, 4- and 6-fold axes are allowed, while 5-, 7- and all higher rotations are disallowed. The proof is very simple, and it is instructive to consider how two parallel 4-fold or 6-fold axes of rotation generate translational symmetry, while two parallel 5-fold axes of rotation clearly cannot coexist.

This proof makes it obvious that 5-fold symmetry is incompatible with translational symmetry, and hence with crystallinity....

....What are quasicrystals?

A quasicrystal is a material that exhibits long-range order in a diffraction experiment and yet does not have translational periodicity. In fact, the assumption that a crystal must be 3-dimensionally periodic had already been challenged by the discovery of incommensurately modulated structures. These are crystal structures that are subject to periodic distortions with a period that is incompatible with that of the underlying parent lattice. The existence of incommensurability was inferred in the structure of cold-worked metals as early as 1927, but a comprehensive treatment in terms of the now-prevalent superspace approach was not introduced until the work of de Wolff and Janner and Jansen. In contrast to quasicrystals, these structures may however be regarded as distortions of periodic structures, and their point-group symmetries allow 3-dimensional periodicity. In lieu of translational periodicity, quasicrystals exhibit another intriguing symmetry property, namely self-similarity by scaling. In icosahedral and decagonal quasicrystals, the self-similarity is related to the scaling properties of the golden ratio τ, $(\sqrt{5} + 1)/2$. This feature is clearly apparent in direct space models and diffraction patterns alike.

The superspace formalism developed to treat incommensurately modulated structures was well adapted to deal also with quasicrystals. Hermann showed that symmetries that are noncrystallographic for 3-dimensional lattices may become crystallographic if treated in higher dimensional space. Icosahedral symmetry is allowed together with translational symmetry in 6-dimensional space, where each coordinate axis is perpendicular to a hyperplane spanned by the other five.

Projection to a 3-dimensional external space is straightforward according to the projection matrix,

$$M = (\tau^2 + 1)^{-\frac{1}{2}} = \begin{pmatrix} 1 & \tau & 0 & -1 & \tau & 0 \\ \tau & 0 & 1 & \tau & 0 & -1 \\ 0 & 1 & \tau & 0 & -1 & \tau \end{pmatrix}$$

The golden ratio τ appears naturally in all manifestations of 5-fold symmetry as the relation between the diagonal and the edge in a regular pentagon, and it is inextricably linked to the Fibonacci sequence. Application of the projection matrix to the vertices of a 6-dimensional hypercube yields a regular icosahedron as the projection. In *the 3+1–dimensional incommensurate example shown, the atomic surfaces are 1-dimensional objects, while in the case of the 6-dimensional hyperspace needed for icosahedral quasicrystals, they are 3-dimensional objects of the appropriate symmetry."....*

During the previous week's tutorial Jack had carefully explained to us that there were a finite number of repeatable arrangements of atoms in a crystal (i.e. 14 possible Miller-Bravais Lattices); and that these led directly to the six possible orders ("families") of rotational crystal symmetry, that is the Triclinic, Monoclinic, Orthorhombic, Tetragonal, Hexagonal and Cubic. Additionally, the three-fold rotational symmetry of the Trigonal sub-system is mathematically subsumed within the six-fold symmetry of the Hexagonal system to give seven rotational symmetry "systems".

Specifically impossible in crystal morphology was the five-fold Pentagonal.

This is Doctrine.

In real life things are of course much more complicated.

The Cube

Very many minerals, and other solid chemicals, exhibit Cubic rotational symmetry. That is to say the repeated atomic structure Unit Cell is a Cube and this implies that if you rotate the crystal about its three orthogonal axes it appears to be identical every ninety degrees (i.e. a full 2π rotation divided by four).

As stated, many common and uncommon minerals present Cubic rotational symmetry, either always or sometimes. Pyrite is a very common mineral indeed. It is the simple Iron Sulfide, FeS_2, and has a lustrous gold color: "Fool's Gold". Confusingly, Pyrite can show as either Cubic crystals or "Pyritohedra", a hybrid shape which is very Phidian but is "really" Cubic at bottom. Just to be even more difficult, Iron Disulfide FeS_2, which is identical to pyrite chemically, can show up in rocks as the mineral Marcasite, often a rosette of silver-bright needles of Orthorhombic crystal form.

This and six other Cubic minerals were key, along with coal, to the pioneering industrialisation of Britain in the eighteenth century: These seven were Pyrite (FeS_2), Halite (common salt) (NaCl), Flourite

(CaF$_2$), Galena (PbS), Sphalerite [(Zn,Fe)S], Copper (Cu) and Magnetite [Fe^{2+}(Fe^{3+})$_2$(O^{2-})$_4$].

In addition Gold (Au), Silver (Ag) and Diamond (C), not to mention numerous less coveted minerals display Cubic symmetry.

But I digress.

We will have further opportunities to discuss pyrite.

I suppose that the cube is arguably the most straight-forward three-dimensional shape.

So it behoves us to take a look at the Cube, and especially to address Pi-Phi relations in and around the Cube.

Figure 14.1 is a quasi-isometric sketch diagram of four cubes standing edge-to-edge or corner-to-corner to illustrate the principle of the packing of Cubic unit cells of atom arrangements in solid chemicals like the mineral ores we have just identified:-

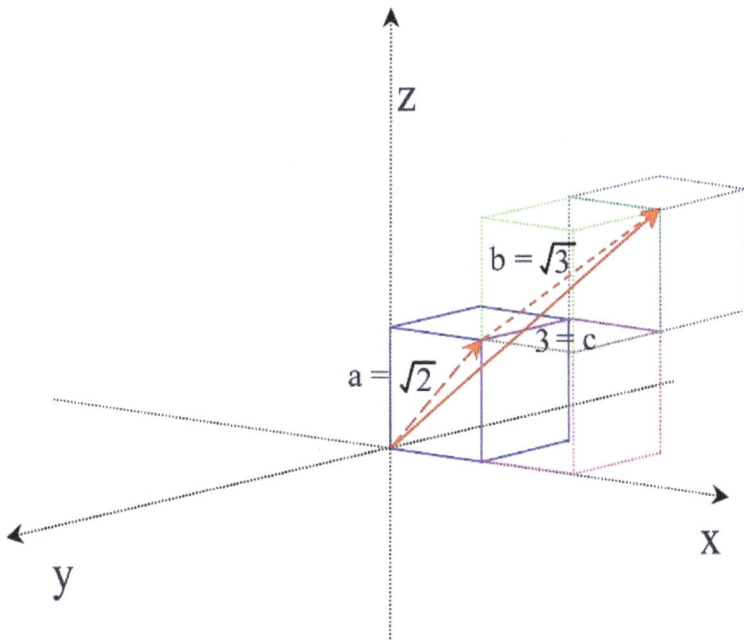

Figure 14.1
The Scheme of Cubic Unit Cell Packing in a Solid

Each of our four cubes has an identical edge length of one unit, and each cube has twelve edges and six sides (including shared elements).

Since the sides and edges of a cube meet orthogonally we can use the Pythagoras Theorem to compute the diagonals that cross the Face or the Body of the cube, and we can use other simple algebra to calculate diagonals *between* cubes.

First of all, recollect the formula for a cartesian Straight Line in Three-Dimensional Euclidian Space:-

$$d_i = \sqrt{(x_{i+1} - x_i)^2 + (y_{i+1} - y_i)^2 + (z_{i+1} - z_i)^2}$$
Equation 14.1

d_i is the Ith Straight-Line Length (distance); and x, y and z are successive Three-dimensional Co-Ordinates along the respective horizontal, depth, and vertical Axes.

To simplify this exercise we can assume that for any cube in question x_i, y_i and $z_i = 0$, i.e. the origin is the near bottom left-hand corner of the cube; and that the Cube Edge Length a is unity..

This leads immediately to:-

$$d_i = \sqrt{(x)^2 + (y)^2 + (z)^2}$$
Equation 14.2

where x = y = z =1 as aforestated.

Allow that d_{face} is the Diagonal Distance along the Face of a Cube and d_{body} is the Diagonal Distance through the Body of a Cube. Then:-

$$d_{face} = \sqrt{x^2 + z^2} = \sqrt{1^2 + 1^2} = \sqrt{2} = a$$
Equation 14.3

whilst:-

$$d_{body} = \sqrt{x^2 + y^2 + z^2} = \sqrt{1^2 + (-1)^2 + 1^2} = \sqrt{3} = b$$
Equation 14.4

We may re-employ Equation 14.4 to compute the multi-cube diagonal c transiting the three blue, purple and green cubes between points (0,0,0) and (2,1,2):-

$$d_{multi} = \sqrt{x^2 + y^2 + z^2} = \sqrt{2^2 + (-1)^2 + 2^2} = \sqrt{9} = 3 = c$$
Equation 14.5

Do please note that the angle (0,0,0):(1,0,1):(2,-1,2) is *not* 3π/4.

But we can apply the Cosine Rule to calculate the size of this angle, that we will call θ:-

$$\theta = \cos^{-1}\left(\frac{c^2 - a^2 - b^2}{-2ab}\right) = \cos^{-1}\left(\frac{9 - 2 - 3}{-2\sqrt{2}\sqrt{3}}\right)$$
Equation 14.6

or:-

$$\theta = \cos^{-1}\left(-2^{\frac{1}{2}}.3^{-\frac{1}{2}}\right) = \cos^{-1}\left(\sqrt{\frac{2}{3}}\right) \approx \phi$$

Equation 14.7

θ has a value of about 0.615479708670387 radians or approximately 144.73561°. The PSD(φ,θ) = 0.413291198543015 representing an error of about 1 part in 242.

Heronian Area of the Triangle Subtending θ

Note that:-

$$\pi \approx \sqrt{2} + \sqrt{3} = d_{face} + d_{body}$$
Equation 14.8

The value of $2^{\frac{1}{2}} + 3^{\frac{1}{2}}$ is about 3.14626436994197 and accordingly PSD(π,$2^{\frac{1}{2}} + 3^{\frac{1}{2}}$) = -0.148705350034519

Heron's Law for the computation of the Area, A, of a triangle given its three side lengths a, b and c may, via the Semi-Perimeter, s, be specified in these terms:-

$$s = \frac{a+b+c}{2}$$

Equation 14.9

and hence:-

$$A = \sqrt{s(s-a)(s-b)(s-c)}$$

Equation 14.10

Noting *inter alia* that:-

$$s = \frac{a+b+c}{2} = \frac{\sqrt{2}+\sqrt{3}+3}{2} \approx \frac{\pi+3}{2}$$

Equation 14.11

we can make the appropriate substitutions to obtain:-

$$A = \sqrt{\frac{\sqrt{2}+\sqrt{3}+3}{2}\left(\frac{\sqrt{2}+\sqrt{3}+3}{2}-\sqrt{2}\right)\left(\frac{\sqrt{2}+\sqrt{3}+3}{2}-\sqrt{3}\right)\left(\frac{\sqrt{2}+\sqrt{3}+3}{2}-3\right)}$$

Equation 14.12

or:-

$$A = \sqrt{\frac{1}{16}(\sqrt{2}+\sqrt{3}+3)(\sqrt{3}-\sqrt{2}+3)(\sqrt{2}-\sqrt{3}+3)(\sqrt{2}+\sqrt{3}-3)}$$

Equation 14.13

which may be re-arranged as:-

$$A = \sqrt{\frac{1}{16}\left[2\sqrt{2}^2\left(\sqrt{3}^2 + 3^2\right) - \left(\sqrt{3}^2 - 3^2\right)^2 - \sqrt{2}^4\right]}$$

Equation 14.14

This highly-redundant expression condenses to:-

$$A = \sqrt{\frac{1}{2}} = \frac{1}{\sqrt{2}} = 0.707106781186547$$

Equation 14.15

Use of the "Pi" Approximation

We may state the following approximants:-

$$\pi \approx \sqrt{2} + \sqrt{3}$$

Equation 14.16a

$$\frac{1}{\pi} \approx \sqrt{3} - \sqrt{2}$$

Equation 14.16b

$$-\frac{1}{\pi} \approx \sqrt{2} - \sqrt{3}$$

Equation 14.16c

As we have seen PSD(π,$2^{\frac{1}{2}}$+$3^{\frac{1}{2}}$) = -0.148705350034519, and the PSD for the two reciprocal expressions is +0.148484545571338

It follows by substitution that the Approximate Area, A_{approx}, of the trans-cubic triangle is given by:-

$$A_{approx} = \sqrt{\frac{1}{16}(\pi + 3)\left(\frac{1}{\pi} + 3\right)\left(-\frac{1}{\pi} + 3\right)(\pi - 3)}$$

Equation 14.17

The bracketed product is a transcendental number.

Equation 14.17 simplifies in the following manner:-

$$A_{approx} = \sqrt{\frac{1}{16}\left(9\pi^2 + \frac{9}{\pi^2} - 82\right)}$$

$$= \sqrt{\frac{1}{16}\left[9\pi^2 + \frac{9}{\pi^2} - (9^2 + 1)\right]}$$

$$= \sqrt{\frac{9}{16}\left[\pi^2 + \frac{1}{\pi^2} - \frac{9^2 + 1}{9}\right]}$$

$$= \sqrt{\frac{9}{16}\left[\pi^2 + \frac{1}{\pi^2} - \left(9 + \frac{1}{9}\right)\right]}$$

Equation 14.18

and hence:-

$$A_{approx} = \frac{3}{4}\sqrt{\left[\left(\pi^2 + \frac{1}{\pi^2}\right) - \left(9 + \frac{1}{9}\right)\right]}$$

Equation 14.19

The value of A_{approx} is 0.695446361275677 and the PSD(A,A_{approx}) is 1.6490323980919 percent. Thus this elegant outcome is seriously inaccurate.

As a mitigant we may transform the expression beneath the square-root sign into a quadratic equation employing the factor κ and extend Equation 14.19 as:-

$$A_{approx} = \frac{3}{4}\sqrt{\left[\left(\pi^2 + \frac{1}{\pi^2}\right) + \kappa\pi - \left(9 + \frac{1}{9}\right)\right]}$$

Equation 14.20

Re-arrangement for an explicit quadratic permits us to write:-

$$A_{approx} = \frac{3}{4}\sqrt{\left[\pi^2\left(1+\frac{1}{\pi^4}\right) + \kappa\pi - \left(\frac{82}{9}\right)\right]}$$

Equation 14.21

By setting A_{approx} to $1/(2^{\frac{1}{2}})$ and simplifying the resulting equations we may establish that:-

$$\kappa = \frac{1}{\pi}\left(10 - \frac{1+\pi^4}{\pi^2}\right)$$

Equation 14.22

This allows us to specify:-

$$A_{approx} = \frac{3}{4}\sqrt{\left[\pi^2\left(1+\frac{1}{\pi^4}\right) + \pi\frac{1}{\pi}\left(10 - \frac{1+\pi^4}{\pi^2}\right) - \left(\frac{82}{9}\right)\right]}$$

Equation 14.23

Equation 14.23 simplifies to:-

$$A_{approx} = \frac{3}{4}\sqrt{\left[\pi^2\left(1+\frac{1}{\pi^4}\right) + \left(\frac{8}{9} - \frac{1+\pi^4}{\pi^2}\right)\right]}$$

Equation 14.24

Equation 14.24 has a Percentage Specific Defect relative to the analytic Area A given as:-

$$PSD\left(A, A_{approx}\right) = 0.000000000000094$$

Equation 14.25

We are entitled to think that the discrepancy is numerical error and that A and A_{approx} are analytically equivalent.

Equation 14.24 immediately simplifies to:-

$$A_{Approx} = \frac{3}{4}\sqrt{\frac{8}{9}} = \frac{3}{4} \cdot \frac{2\sqrt{2}}{3} = \frac{2\sqrt{2}}{4} = \frac{\sqrt{2}}{2} \equiv \frac{1}{\sqrt{2}}$$

Equation 14.26

The trans-cubic area is therefore a function neither of π nor of Φ.

But I am aware that we are supposed to be discussing minerals in this chapter, and that accordingly I should as a geologist come back to earth, so to say.

Pyrite

Earlier, we briefly alluded to Pyrite (or Pyrites, sometimes called Mundic or Fool's Gold). It is chemically Iron Disulfide (FeS_2), though this chemical can appear in a different crystal form as Marcasite. Both types are insoluble in water and common in sediments that deposited in anaerobic, chemically-reducing conditions.

Such conditions were widespread in Palaeozoic muds, and especially in the coals that formed in the Upper Carboniferous (vaguely Westphalian or Pennsylvanian) from the induration of paludinal tree-like vegetation.

Therefore, pyrite often forms amorphous looking veins and laminae in British coal (a pollutant called brazzle). Pyrite in coal is highly undesirable industrially: At concentrations of 5-7% it increases the risk of spontaneous combustion in coal-heaps; it causes pneumoconiosis in those exposed to pyritic coal; and it compounds the difficulty and expense of steel production.

Where however pyrite precipitates from a fluid state in vesicles and vuggs, for example in pre-existing voids such as solution cavities, it has the liberty to develop lovely and remarkable brassy crystals as definite single or interpenetrating cubes or as "pyritohedra".

Photo: James R Warren

Figure 14.2
Interpenetrating Pyrite Natural Cubes
From the Victoria Mine, Navajún, La Rioja, Spain

Figure 14.2 shows a pyrite crystal specimen from the author's cabinet. I purchased this at Lyme Regis, but it is almost certainly from Victoria Mine at Navajún, and it sits in a chloritic Chamosite $(Fe^{2+},Mg,Al,Fe^{3+})_6(Si,Al)_4O_{10}(OH,O)_8$ ("marl") matrix.

Mineral crystals exhibit a characteristic crystal system property that manifests at the macroscopic level as consistency of interangles between faces and edges, but is controlled by the underlying structure of atomic arrangements.

In addition, the visible crystal often exhibits "habit", a superficially distinct solid geometry which is conditioned by the genesis of the mineral deposit. This outcome is the effect of the assemblage of thermodynamic and other physico-chemical conditions bearing upon the crystal during its formation.

"Pyritohedral" Habit

A "pyritohedron" is a species of irregular dodecahedron whose pentagonal sides can vary and are thus potentially irregular themselves. A strict mathematical pyritohedron has twelve identical irregular pentagonal sides.

The ideal pyritohedron has twenty vertices and thirty edges. Twenty-four of the edges are of one length and the remaining six a different length.

Figures 14.3 and 14.4 represent the ideal of the pyritohedron.

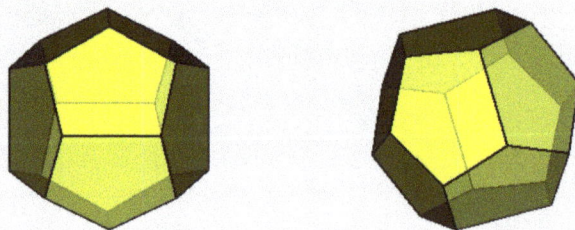

Computational Simulagraphs from Minerals.Net Minerals Glossary

Figures 14.3 and 14.4
Aspects of a Mathematically Ideal Pyritohedron

Figure 14.5 is a photograph of a rare and incredibly beautiful, but not mathematically perfect, natural pyritohedral pyrite crystal from Shangbao Pyrite Mine in Hunan Province, China:-

Photo: Robert M Lavinsky

Figure 14.5
Pyritohedral Pyrite from Hunan in China

In my photograph, and especially those of Professor Lavinsky, the striations characteristic of pyrite faces are well displayed. Some researchers say that these striations contain more crystallographic information about Phi than even the crystal face[14.2,14.3,14.4].

Professor Lavinsky further presents the Arizonan specimen Pyrite-193871 which presents hackly but metrically-useful faces. This is shown as Figure 14.6. Contributor Cityscape most usefully moved forward to annotate the frontal face with face internal angles 121.6°-106.6°-102.6°-102.6°-106.6° which total 540° or 3π radians as they must for any pentagon. His presentation is reproduced as Figure 14.7.

We may recollect that the Golden Angle, γ, is given by:-

$$\gamma = \pi\left(3 - \sqrt{5}\right)$$
Equation 14.27

and accordingly:-

$$\frac{1}{\Phi^2} \cdot \frac{3\pi}{\gamma} = \frac{3}{2}$$
Equation 14.28

Photo: Robert M Lavinsky

Figure 14.6
Pyritohedral Pyrite-193871 from Arizona

Photo: Citynoise

Figure 14.7
Annotated
Pyritohedral Pyrite-193871 from Arizona

I have summarised some of these facts in Table 14.1:-

Major Ratio of Phidias	Symbol	Value	Radians	Degrees
	Φ	1.618034		
Golden Angle		2.399963	2.399963	137.5078

Diagram

	Value
Long Side (mm)	211.56
Width	282.08
Theory Height	236.5313
Measured Height	235.59
Theory Short Side	161.5816
Measured Short Side	161
PSD(T$_{Height}$,M$_{Height}$)	0.397948
PSD(T$_{ShSide}$,M$_{ShSide}$)	0.359954

Degrees	Radians	Fraction of Φ	Fraction of Golden Angle	Number of Angles in Face	Sub Total (Degrees)	Sub Total (Radians)
102.6	1.790708	1.106718	0.746140	2	205.2	3.58141563
106.6	1.860521	1.149865	0.775229	2	213.2	3.72104197
121.6	2.12232	1.311666	0.884314	1	121.6	2.12232037
Totals 330.8	5.773549	3.56825	2.405682	5	540	9.42477796
3π Radians						9.42477796

Table 14.1
The Irregular Pentagonal Face of the Pyritohedron Dodecahedron

The Surface Area of the Pyritohedron as an Irregular Dodecahedron[14.5]

In terms of the side portrayed in Figure 14.7 and as per the Wikipedia article "Dodecahedron" we may define the face geometry in terms of Figure 14.8:-

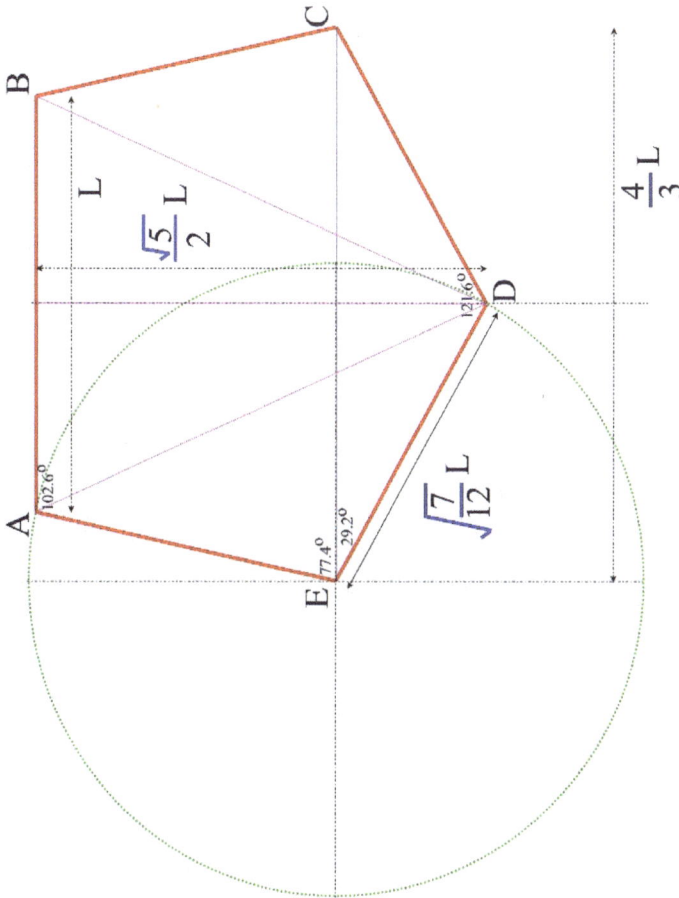

Figure 14.8
The Angles and Edges of a Pyritohedron Face

L is the Long Edge AB = unity, whilst d is the Diagonal AD = BD. From these it follows that Height, H, is given by:-

$$H = \frac{\sqrt{5}}{2}L$$
Equation 14.29

and that that a Short Edge, s = BC =CD = DE = EA is given by:-

$$s = \sqrt{\frac{7}{12}}L$$
Equation 14.30

Accordingly, the Perimeter, P, of the Face is:-

$$P = L + 4s = L + 4\sqrt{\frac{7}{12}}L = L\left(1 + 4\sqrt{\frac{7}{12}}\right)$$
Equation 14.31

The Area of the Face, A, is:-

$$A = 2.AED + ABD$$
Equation 14.32

and hence the Surface Area, S, of the Pyritohedron is:-

$$S = 12A$$
Equation 14.33

Angles

With regard to the facial angles:-

$$\alpha = 121.6° = 2.1223203704251 \ radians$$
Equation 14.34a

$$\beta = 106.6° = 1.86052098262596 \ radians$$
Equation 14.34b

$$\gamma = 102.6° = 1.79070781254618 \; radians$$
Equation 14.34c

Please note that this γ is *not* the Golden Angle: I am guilty of symbolic re-cycling.

Therefore the Sum of the Internal Angles of the Face, θ, is given by:-

$$\theta = \alpha + 2(\beta + \gamma) = 3\pi = 9.42477796076938 \; radians$$
Equation 14.35

Angular Ratios

$$\alpha_\pi = \frac{\alpha}{\pi} = 0.675555555555556$$
Equation 14.36a

$$\beta_\pi = \frac{\beta}{\pi} = 0.592222222222222$$
Equation 14.36b

$$\gamma_\pi = \frac{\gamma}{\pi} = 0.57$$
Equation 14.36c

And it follows that:-

$$\frac{\alpha + \beta + \gamma}{\pi} = 1.83777777777778$$
Equation 14.37

and:-

$$\frac{\alpha + 2\beta + 2\gamma}{\pi} = 3$$
Equation 14.38

The Value of Diameter d

By the Cosine Rule:-

$$d = \sqrt{s^2 + s^2 - 2ss.\cos(\beta)}$$
Equation 14.39

therefore:-

$$d = \sqrt{2s^2[1 - \cos(\beta)]} = 1.22473252670633$$
Equation 14.40

Moving forward by substitution for s:-

$$d = \sqrt{\sqrt{\frac{7}{12}}L^2 + \sqrt{\frac{7}{12}}L^2 - 2.\sqrt{\frac{7}{12}}L\sqrt{\frac{7}{12}}L.\cos(\beta)}$$
Equation 14.41

so:-

$$d = \sqrt{\frac{7}{6}L^2(1 - \cos(\beta))}$$
Equation 14.42

The Face Perimeter, P

This is given by:-

$$P = L + 4\left(\sqrt{\frac{7}{12}}L\right) = L\left(1 + 4\sqrt{\frac{7}{12}}\right) = 4.05505046330389$$
Equation 14.43

given that L is unity.

The Face Area, A

As aforenoted:-

$$A = 2.AED + ABD$$
Equation 14.44

The area of the component triangle AED (please see Figure 14.8) is yielded by:-

$$AED = A_{AED} = \frac{1}{2}s^2.\sin(\beta) = 0.278510750885664$$
Equation 14.45

whilst that of triangle BCD is given by:-

$$BCD = A_{BCD} = \frac{1}{2}s^2.\sin(\beta) = 0.278510750885664$$
Equation 14.46

AED and BCD are identical and may be grouped together as 2×(AED+BCD) for the purpose in hand.
The angle ψ is:-

$$\psi = \frac{\pi - \angle AED}{2} = 36.7° = 0.640535835481919$$
Equation 14.47

and hence Area ABD or A_{ABD} is :-

$$ABD = A_{ABD} = \frac{1}{2}d^2\sin(\alpha - 2\psi) = \frac{HL}{2} = 0.559095728900246$$
Equation 14.48

Accordingly:-

$$A_{face} = A_{AED} + A_{BCD} + A_{ABD} = 2A_{AED} + A_{ABD} = 1.11803849614628$$
Equation 14.49

and we may continue with:-

$$A_{face} = \frac{7}{12}L^2 \sin(\beta) + \frac{\sqrt{5}}{2}\frac{L^2}{2}$$

$$= \frac{L^2}{4}\left[\frac{7}{3}\sin(\beta) + \sqrt{5}\right]$$

$$= \frac{L^2}{4}\left[\frac{7}{3}\sin(\beta) + (2\Phi + 1)\right]$$

Equation 14.50

Noting that:-

$$\frac{7}{3}\sin(\beta) \approx \Phi^3 - 2$$

Equation 14.51

with a Percentage Specific Defect given by:-

$$PSD\left(\frac{7}{3}\sin(\beta), \Phi^3 - 2\right) = 0.000806301075375$$
Equation 14.52

we may, assuming equality, continue with the following substitutions:-

$$A_{face} = \frac{L^2}{4}[(\Phi^3 - 2) + (2\Phi - 1)]$$
Equation 14.53

Equation 14.53 may be rewritten as:-

$$A_{face} = \frac{L^2}{4}[\Phi - 1][(\Phi(\Phi + 1) + 3]$$
Equation 14.54

or in terms of the Minor Ratio of Phidias, ϕ, as:-

$$A_{face} = \frac{L^2}{4}(\phi^3 + 3\phi^2 + 5\phi)$$
Equation 14.55

Alternative Expressions of Pyritohedron Surface Area, S

In the light of the foregoing it is instructive to compare alternative computations of the Pyritohedron Surface Area, S, by our three distinct avenues:-

Firstly, S may be simply stated as twelve times the individual Face Area A_{face}:-

$$S1 = 12A_{face} = 13.4164078649987$$
Equation 14.56a

$$S2 = (3L^2)\left(\frac{7}{3}\sin(\beta) + \sqrt{5}\right) = 13.4164619537553$$
Equation 14.56b

$$S3 = (3L^2)(\phi^3 + 3\phi^2 + 5\phi) = 13.4164078649987$$
Equation 14.56c

The following Percentage Specific Defects were computed to pertain:

$$PSD(S1, S2) = -0.000403153788314$$
Equation 14.57a

$$PSD(S1, S3) = 0$$
Equation 14.57b

$$PSD(S2, S3) = -0.000403152162991$$
Equation 14.57b

I cannot explain the discrepancies involving S2 but Sine and Square Root are both functions solved imprecisely using computational algorithms and they could conceivably have led to factitious errors, as possibility borne out by the near symmetry of PSD(S1,S2) and PSD(S2,S3).

The Quasicrystal[14.6]

The French were ever the best mathematicians, at least since the sixteenth-century reforms of symbolism pioneered by the likes of François Viète, Seigneur de la Bigotière.

During the course of the eighteenth century the more "rational" aspects of minerals and other chemicals became the subject of "classical" scientific crystallography at the hands of René Just Haüy, the Abbé Haüy; and his contemporaries, men like Romé de L'Isle. Haüy noted, following the 1669AD work of Nicolaus Steno, that the interfacial angles and the angles between cleavage planes (if any) within crystals were of consistent angular values, as measured with an instrument called a goniometer. If the crystals were smashed, the fragments conserved such angles however tiny the fragments. This led Haüy to postulate the indivisible cells which he called *molé constituantes*; "integrant molecules" or as we would say Unit Cells: This Unit Cell is essentially a geometrical atomic arrangement characteristic of the chemical, and theoretically infinitely propagable without changing the chemical formula. The abbé published his work as *Traité de minéralogie* (1801). René Just Haüy appreciated the taxonomical potential of his work but the working-out of classical crystallography, and its applied science mineralogy had to await the nineteenth-century developments in atomic theory.[14.7]

Later, in 1839, British mineralogist and crystallographer William Hallowes Miller was able to show that the distances along the edges of Unit Cells, bore integer ratios to one another and that particular planes could thus be assigned cartesian-style *relative* integer co-ordinates as $\langle h,k,l \rangle$. This enabled crystal surfaces to be specified consistently and unambiguously to the benefit of the mineralogist and the mathematician. There was no imputation that the crystals or their Cells were necessarily cartesian cubes: Merely that they were parallelepipeds with these *relative* edge ratios.

Neither Miller nor Abbé Haüy had any idea about the *absolute* lengths of these edges in unit cells. Such knowledge had to await the development of x-rays at the end of the century, and because the wavelength of x-rays approximated the distances between atoms in crystals, it became practical for *absolute* distances to be inferred from simple analytic geometry as applied to x-ray diffraction experiments.

In the early years of the twentieth-century, William Henry Bragg and his son Sir William Lawrence Bragg performed the required experiments which output results in the form of patterned dots in two-dimensions on a photographic plate.

An upshot of this stage of science was that *absolute* distances in angstroms or picometers or whatever could be calculated between the various Millerian planes with a crystal. This meant that Lattice Constants could be computed using the equation:-

$$a = d_{h,k,l} \times \sqrt{h^2 + k^2 + l^2}$$
Equation 14.58

These Lattice Constants could be calculated and cataloged uniquely to identify a mineral species.

Furthermore, such knowledge led directly to V, the Volume of the Unit Cell and thus to Interatomic ("Bond") Distances.

None of this violated the "laws" of Classical Crystallographic Science. In fact, predictably enough, theoreticians used this material to "confirm" (I initially typed "conform": It must have been a Freudian slip) the received Doctrine.

A quasicrystal is unobvious. Quasicrystals like to hide in crystalline metallic alloys, especially those on Al-Fe-Cr-Cu-Mn-Pd systems such as $Al_{62.2}Cu_{25.3}Fe_{12.5}$ (note the non-integral atom counts). Quasicrystals are aperiodic. They can be every bit as lovely as classical crystals as the following photograph demonstrates[14.8].

The Holmium-Magnesium-Zinc crystal shown is (despite pitting) a virtually perfect classical dodecahedron with regular pentagonal faces. Accordingly it is very rich in Phidian dimensions, even discounting those that characterise the 2D pentagonal faces. A Platonic Dodecahedron has twelve equal faces, twenty equally-distant vertices and thirty equal edges.

Photo: US Department of Energy

Figure 14.9
A Nearly Perfect Platonic Dodecahedron
as a Ho-Mg-Zn Crystal

Allow that the Regular Dodecahedra Edge Length a is for our purposes unity, and the r_u is the Vertex Radius of the Dodecahedron from its Centroid to any of its twenty vertices; r_i is the Inscribed Sphere Radius tantamount to a radius from the Centroid perpendicular to any Face; r_m is the Mid-radius that connects the Centroid to the Mid-point of any Edge; A is the Dodecahedron Surface Area and V is the Dodecahedron Volume.

If this Ho-Mg-Zn quasicrystal is indeed a Platonic dodecahedron then:-

$$r_u = a \frac{\sqrt{3}}{4} (1 + \sqrt{5}) = a \frac{\sqrt{3}}{2} \Phi$$

Equation 14.59

$$r_i = a \frac{1}{2} \sqrt{\frac{5}{2} + \frac{11}{10} \sqrt{5}} = a \frac{\Phi^2}{2\sqrt{3 - \Phi}}$$

Equation 14.60

$$r_m = a \frac{1}{4} (3 + \sqrt{5}) = a \frac{\Phi^2}{2}$$

Equation 14.61

Furthermore:-

$$A = a^2 . 3 \sqrt{25 + 10\sqrt{5}} = \frac{15\Phi^2}{\sqrt{3 - \Phi}}$$

Equation 14.62

and:-

$$V = a^3 . \frac{1}{4} (15 + 7\sqrt{5}) = \frac{5\Phi^3}{6 - 2\Phi}$$

Equation 14.63

If S_D is the Surface Area of the Dodecahedron and S_S is the Surface Area of its Circumsphere then we may develop the Ratio of the Sphere Area to the Dodecahedron Area, ρ_S, as:-

$$\rho_S = \frac{S_S}{S_D} = \frac{4\pi r_u^2}{a^2 . 3 \sqrt{25 + 10\sqrt{5}}} = \frac{\pi \Phi^2}{\sqrt{25 + 10\sqrt{5}}}$$

Equation 14.64

Similarly, if V_D is the Volume of the Dodecahedron and V_S is the Volume of its Circumsphere then the Ratio of the Sphere Volume to the Dodecahedron Volume, ρ_V, is given by:-

$$\rho_V = \frac{V_S}{V_D} = \frac{\frac{4}{3}\pi r_u{}^3}{a^3 \cdot \frac{1}{4}\left(15 + 7\sqrt{5}\right)} = \frac{2\pi\sqrt{3}\Phi^3}{15 + 7\sqrt{5}}$$

Equation 14.65

In 2016 Bindi, Lin, Chi Ma and Steinhardt[14.9] reported their discovery of icosahedral natural quasicrystals of nominal formula $Al_{62.0(8)}Cu_{31.2(8)}Fe_{6.8(4)}$ in a chondritic (rocky rather than metallic) meteorite that was retrieved from the vicinity of the river Khatyrka in the Arctic extreme North-East of the Russian Federation.[14.10]

The icosahedron is the final and ultimate regular Platonic solid. It is a figure with twelve vertices at equal distances from its centroid; it has thirty equal edges and twenty equal triangular faces.

It is very rich in Phidian dimensions and as with the dodecahedron its geometry can associate π and Φ via relations between the icosahedron and its circumcircle.

The icosahedron is illustrated in Figure 14.10

Full geometric details are available elsewhere, including the Wikipedia[14.12] article "Regular Icosahedron" but we may note in passing that the Vertex Radius (and of course that of the circumsphere), r_u, is given by:-

$$r_u = \frac{a}{2}\sqrt{\Phi\sqrt{5}} = \frac{a}{4}\sqrt{10 + 2\sqrt{5}} = a\sin\left(\frac{2\pi}{5}\right)$$

Equation 14.66

This leads to a Form Filling Factor, $f \equiv 1/\rho_V$, given by:-

$$f \equiv \frac{1}{\rho_V} = \frac{V_I}{\frac{4}{3}\pi r_u{}^3} = \frac{20\left(3 + \sqrt{5}\right)}{\pi\left(2\sqrt{5} + 10\right)^{\frac{3}{2}}} \approx 0.6054613829$$

Equation 14.67

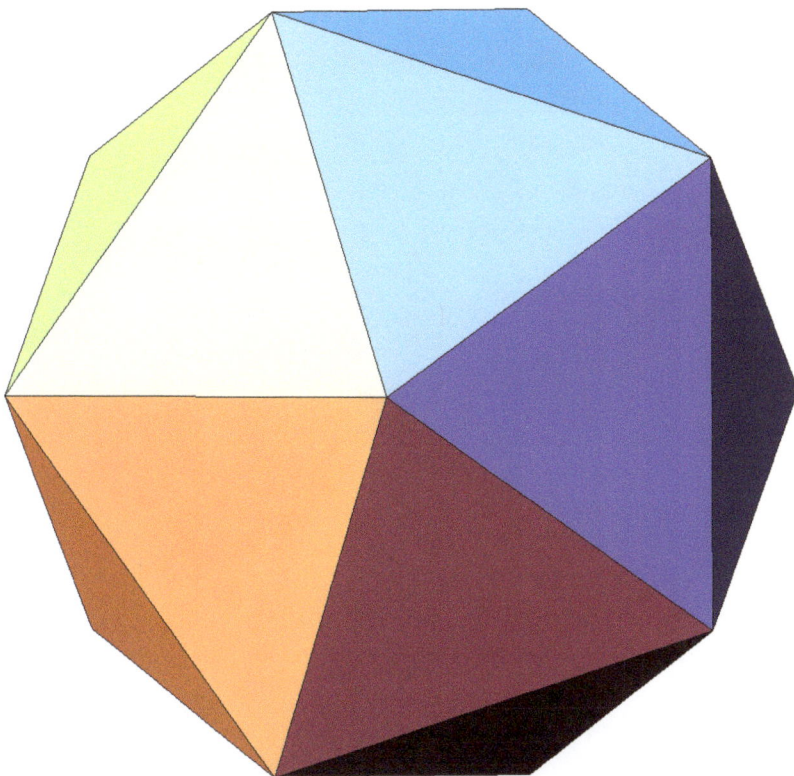

Computational Simulagraph by Tomruen using the Bulatov applet[14.11]

Figure 14.10
A Perfect Regular (Platonic) Icosahedron

In this brief discussion of Pi and Phi in the Mineral Kingdom we have seen how that odd couple of infinitely-deep numbers inhabit quite classical simple minerals like Pyrite and come back with redoubled permeation in the exotic quasicrystals, artificial and natural.

Our arguments could of course be applied to other structured chemicals, and even structure generally.

By the way, I never did get to the bottom of Jack's book, neither in more than fifty subsequent years did I solve the mystery of his golden pyramid.

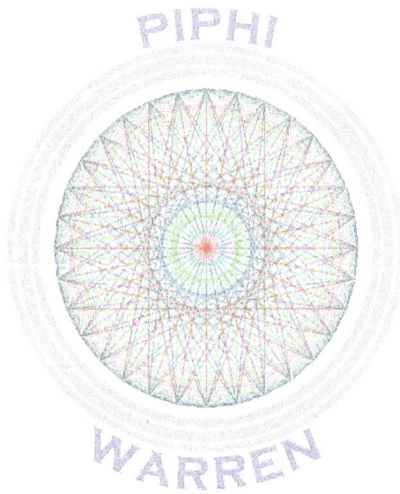

CHAPTER FIFTEEN
PHI IN ART

Many over the centuries have thought that they have detected traces (literal or metaphorical) of the use of the Phidian Ratios in art and architecture, or even in music.

Indeed, Phidias himself was assumed by some to have based the design of The Parthenon upon the Golden Ratio of 1:1.618 approximately, and rumored to have used the same proportion in his planning of statues of the human form, though to the best of my knowledge and belief, no incontrovertible specimens of such works of his hand have been positively identified.

When art works allegedly embedding Phidian proportions have been studied by aesthetes and historians three or four major motives for Phidian planning have been surmised:-

A. Workmanlike Picture Planning

The artist has used (to him) quick and simple geometric constructions, perhaps with ruler and compass, to lay-out convenient zones of operation upon the naked substrate, be it canvas, dimension stone or indeed ground plot.

This compartmentalisation possibly assisted his work program, especially in respect of delegations to contractors or apprentices.

The English proto-impressionist painter JMW Turner is known to have worked in this way, and other contract painters including Leonardo da Vinci and Nicolas Poussin are suspected.

B   Aesthetic Considerations

Many harbor the prejudice that the Golden Ratio $1:\Phi$ defines the most beautiful rectangle and occasionally psychologists have attempted to prove the point by surveying preferences in statistical populations.

Some late nineteenth-century French Impressionists and several twentieth-century artists consciously used the Ratio of Phidias on this assumption.

It is difficult to justify such selection, especially as opposed to the proportions 3:5 and 3:4, and of the two hundred or so national flags, only one is definitely influenced by Φ.

C   Arcane Symbolism

Some authorities allege that old master paintings of the Renaissance and the seventeenth-century include hidden messages, usually by the positioning of significant elements, which may or may not be based in or about geometrical substructures such as pentagrams or other Φ-related mathematical objects..

Such thinking is usually associated with recognition of the extreme peril which attended European heterodoxy and the need to imply, but not to state, heretical opinions: Sometimes at the insistence of the paying patrons, even clerics.

In this context, Nicolas Poussin and his imitators are often studied in art historical criticism.

In addition to these popular currents of interpretive thought we may, in the context of architecture and engineering, add:-

D   Structural Stability Theory

Orthostatic architectonic forms, such as the column-and-lintel porticos, stoas and colonnades typical of Ancient Greece, involve placing a stone beam between two tall supports.

The beam is subject to tensile bending moments tending to tear its lower edge, whilst dispersive moments affect both the tops of the columns and the ends of the beam.

The proportions of the gap between the columns, and the columns' height, can be influenced by scientific theory or empirical findings about the integrity of brittle members under stress.

A capable artist may, of course, have applied Φ for some or all of these motives.

Templates

When we study the plausibility of Phidian traces in flat work by dead artists interview is, we devoutly hope, impossible.

Therefore, it is convenient for us to erect some convenient templates for comparisons.

An infinite number of Phidian templates are possible, all composed within a Golden Rectangle, though only a few are of other than theoretical merit.

The Vincian Template

The Vincian Template, so styled because we suspect Leonardo da Vinci of employing such a template or something very like it, is a Phidian Rectangle of Φ:1 proportions containing a Golden Spiral and perhaps cross-orthogonals at the Cartesian limits of the volutes. A Vincian template is exampled by Figure 15.1:-

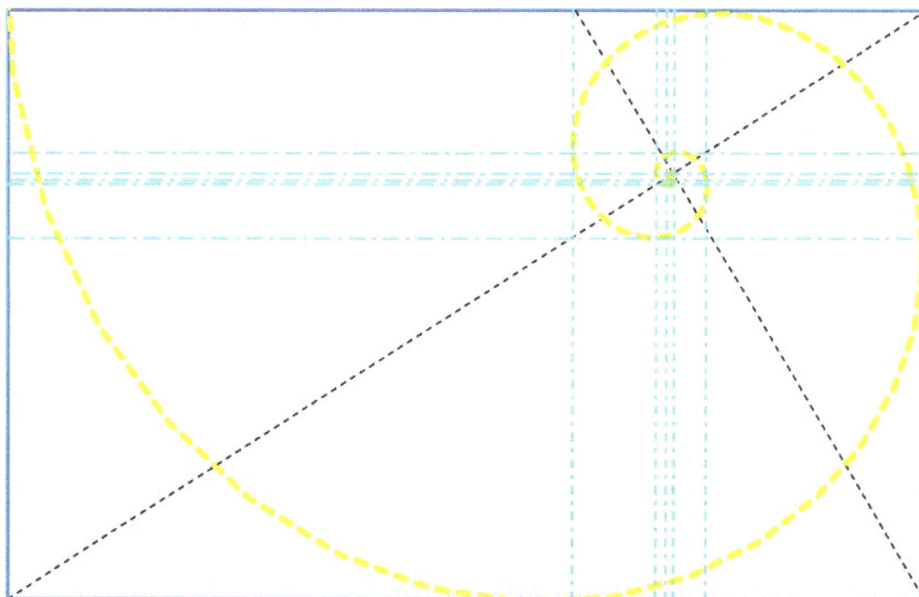

Figure 15.1
A Vincian Template

The light colors of the spiral and the cross-orthogonals, whilst indistinct on a white field, are chosen because they contrast against the dark paint of old canvases.

The Turnerian Template

The Turnerian Template is a guide underlay (or analytical overlay) known to have been used by the English proto-Impressionist Joseph Mallord William Turner during the course of the Early Nineteenth-Century.

This variety of template was discussed by Cook in his 1914 book "The Curves of Life".[15.1]

The Turnerian Template is a Golden Rectangle divided into the exponential proportions Φ^1, Φ^0, Φ^1, Φ^2, Φ^3, Φ^4, Φ^5 along the horizontal axis, with vertical demarcation lines projected above. My example includes horizonal demarcations also.

This results in a quite complex graticule which can lead to over-formalisation of the composition.

I have illustrated a Turnerian Template in Figure 15.2

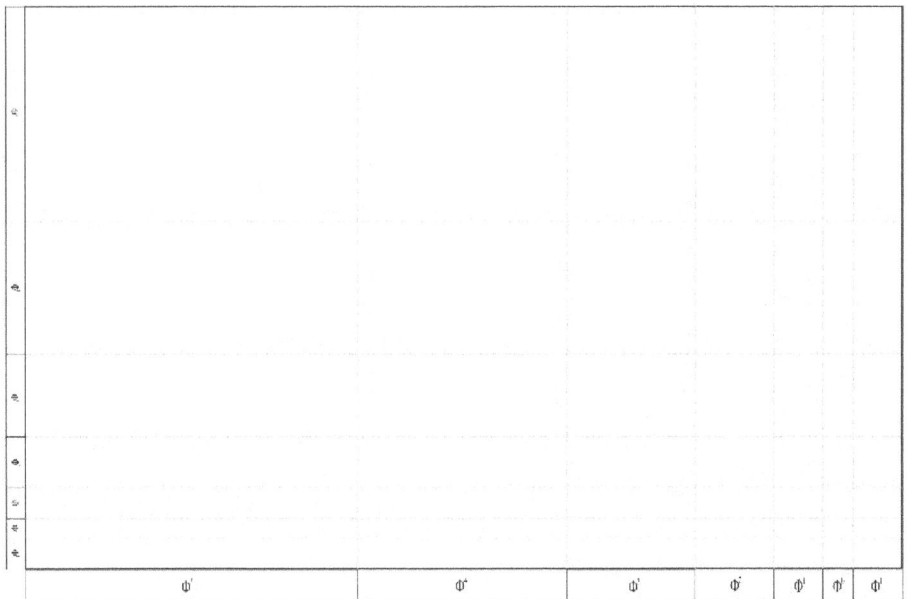

Figure 15.2
A Turnerian Template

Phidian templates can be used not only to plan graphics and artifacts, but also as an adjunct to art historical criticism. They can be employed forensically, so to say, overlayed on existing canvases to determine the geometry (if any) used by the artist in his design. This will later be demonstrated.

When using a template take care not to squeeze or stretch it as illustrated in Figure 15.3. It is, however, permissible to compress or dilate a template proportionately to fit a whole graphic or even just the part relevant to your particular analysis ("affine transformation"). This permissible re-sizing is illustrated in Figure 15.4

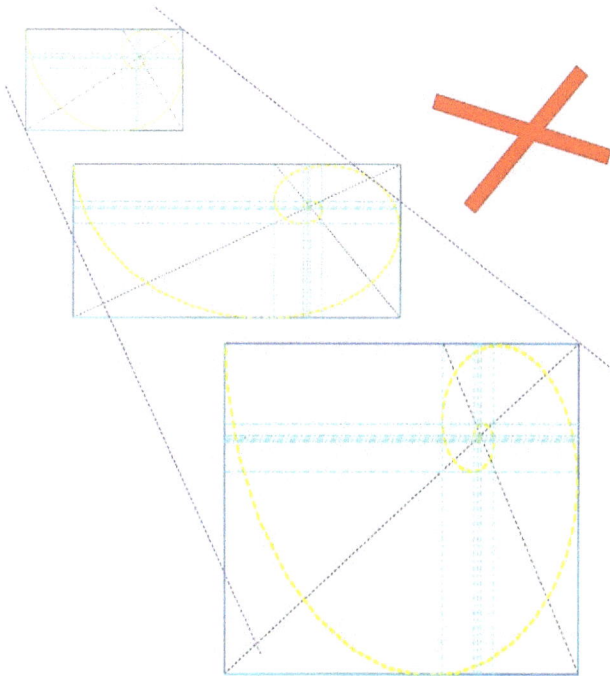

Figure 15.3
Squeezing and Stretching Abuse of a Graphic Template

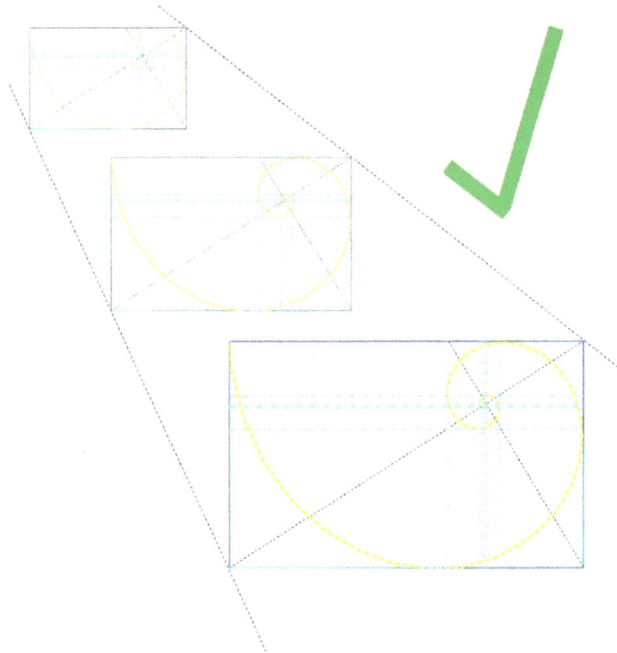

Figure 15.4
Permissible Proportionate Re-Sizing of a
Phidian Graphic Template

Phi and the Higher Nonsense

      Your research will disclose very numerous examples of the abuse of Phidian theory in the arts and sciences, and especially in the analysis and criticism of paintings and plastics by the great Masters of the past.

      Many analysts stretch, squeeze or otherwise warp templates, or even the underlying mathematics itself, to prove a point. The motives are as varied as human corruption. The best of the critics are innocent amateurs, even boys, who tamper with things that have greatly entranced them, but which they are too inexperienced judiciously to adjudge. The worst are those who knowingly prostitute their gifts in quest of money, preferment, fame or the respect of rivals and competitors. Such are too many professors.

      Elsewhere, we discussed a straight-line transect of a stack of cubes and demonstrated that we could construe its algebra as an involved equation masquerading as an elegant function of π and Φ, a length which when unmasked proved to be a simple reciprocal of the Pythagoras Constant. This is not to say such mathematical expansions have no legitimate uses.

The human form and its representation has proven a limitless source of such absurdities.

The Canon of Polykleitos[15.2]

A canon is a set of rules or principles established by Authority, especially the best sort of Authority: Classical Authority.

The Canon of Polykleitos is, however, not a textual list of doctrines in the usual sense: Rather it is a marble sculpture exemplifying the supposedly perfectly-proportioned form in the shape of a divine nude man.

This homunculus, as the Romans would have greeted it, is called The Doryphoros of Polykleitos.

I am not an art historian.

Allow Dr CG Hughes to explain:-

"...Though we do not know the exact details of Polykleitos's formula, because he chose to expound the canon by sculptural rather than discursive means. Although one would expect the free-standing image of the spear-bearer to have been commissioned for a memorial to a deceased warrior, the Doryphoros was not intended to refer to any specific individual, but rather to an idealized composite of all individuals. The Doryphoros is the perfect expression of what the Greeks called symmetria. In art of the High Classical period (ca. 450–400 BC), symmetria not only encompassed a sense of proportion and balance, but was also an exercise in contrasts. The body of the Doryphoros, for example, stands in what is termed contrapposto, meaning that his weight rests on his right leg, freeing his left to bend. In the process, the right hip shifts up and the left down; the left shoulder raises and the right drops. His body is brought into a state of equilibrium through this counter-balancing act.

Although the Doryphoros represents a warrior poised for battle, he does not don a suit of armor or any other protective gear. In fact, were it not for the actual spear that that statue originally held, it would have been difficult to identify him as such. A hallmark of classical Greek sculpture, male nudity or nakedness was understood as a marker of civilization that separated the Greeks from their "barbarian" neighbors.

The Doryphoros was originally executed in bronze, the tensile strength of which allowed for a greater freedom of motion in the statue. The weight of the stone requires the marble copies to have ungainly supports props which diminish the effect intended by Polykleitos."

I reproduce an annotated image of the Doryphoros below:-

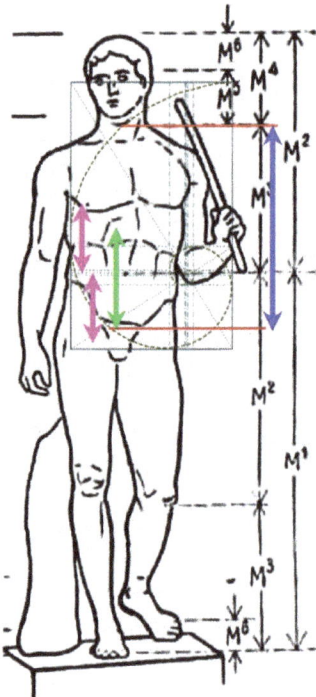

Figure 15.5
The Doryphoros of Polykleitos
Annotated by the Author

In Figure 15.5 I have overlain a feint image of my Vincian template which clearly shows that the model divides the distances from the right eye to the scrotum versus the right eye to the navel in the ratio Φ:1, the navel being at the latitude of the Golden Spiral umbilicus.

The blue arrowed vertical is the distance between the perineum and the laryngeal prominence whilst the green vertical is precisely half that length. The green line approximates the distance between the perineum and the solar plexus. The upper purple line is the distance from the navel to the nipple, equal to the distance from the navel to the perineum.

This contrast between anatomic reality, perfect in The Mind of God, and the aesthetic ideal extends to Vitruvian Man and virtually all artistic conceptions of the human form.

One is compellingly reminded of the late Kenneth Clark's distinction between the naked and the nude.

It is interesting to test the *Doryphoros* against actual naked humans as, for example, the women photographed in the beach scene of Figure 15.6.

The lissome young lady in the center closely approximates the classical ideal of Phidian proportion whereas the template shows that the older women understandably fall short (so to say) in several respects. In the case of the young woman it can be seen that the Polykleitic model is also emulated in the near-perfect equality of distance between the nipple and the navel; and the navel and the perineum.

Whilst these variations are understandable, and acceptable in the broad scheme of things, there are more serious pitfalls that arise in Proofs from Absurdity: These are attempts to prove the truth of Authority by comparing an adopted ideal with similar, but exaggerated, conceptions that we are invited to deride.

An example is shown in Figure 15.7 where the classical ideal (in the center) is contrasted with a figure whose "torso is too long" at the left and a figure whose torso is "too short" at the right. Of course, none of the three figures are natural.

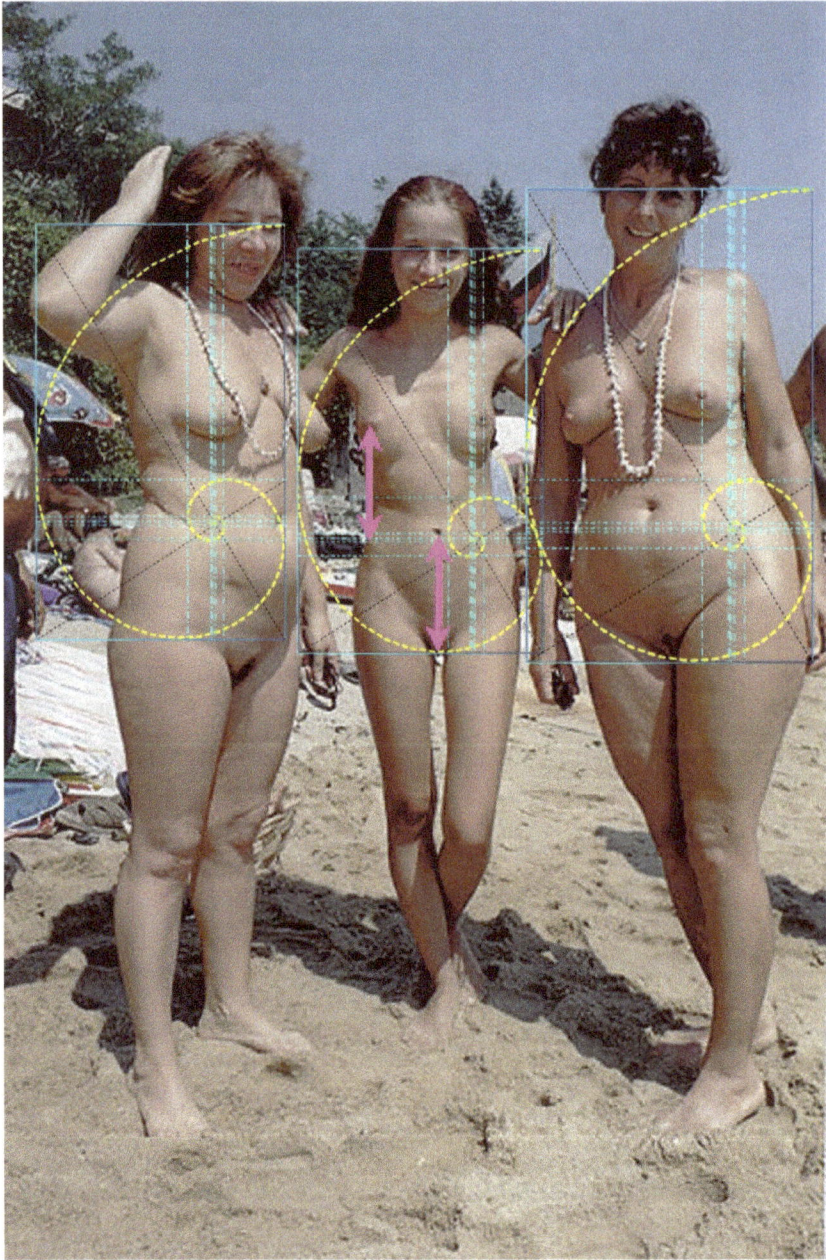

Figure 15.6
The Three Graces of the Beach

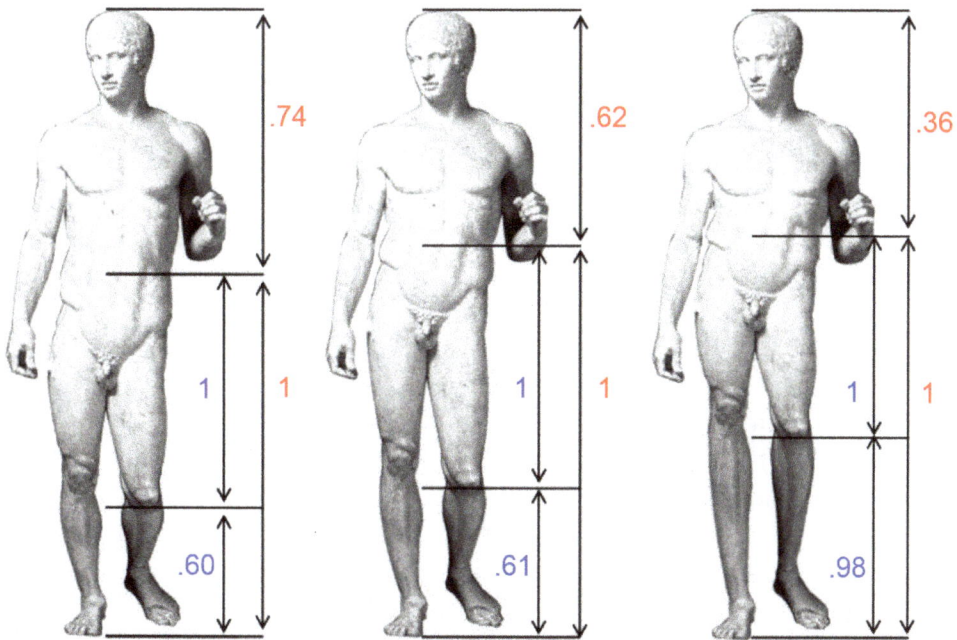

Figure 15.7
Specious Comparison as Aesthetic Advocacy

Suggestive Perspectives

Other fallacies you are likely to find include the abuse of perspective, and the statistically selective, sometimes used to infer that devolving galaxies or meteorological cyclones and so forth are examples of The Golden Spiral.

A typical example is reproduced in Figure 15.8

We may be inclined to accept that devolving galaxies or even the planforms of hurricanes describe *logarithmic* spirals on physical grounds but whilst all Golden Spirals are logarithmic, not all logarithmic spirals are Golden...

Graphical Exemplars:- A. The Jobbing Painter

Earlier in this chapter we briefly identified the Workmanlike Picture Planner and the practical considerations that bore upon his use, real or surmised, of a Phidian template undersketch for the layout of his painting, or plan.

We should wish briefly to study the examples of Leonardo Da Vinci, JMW Turner and Katsushika Hokusai, and maybe others if time allows.

Our first example is *Mona Lisa* or *La Giaconda*, an early sixteenth-century portrait of a Florentine lady by the Tuscan polymathic genius Leonardo da Vinci.

I reproduce it in Figure 15.9, together with my superposition of a Vincian template.

First of all, I am not unduly worried that parts of the canvas seem to have gone missing or even been cut away over the years. Forensic studies often, or even usually, disclose the mutilation of Master paintings over the centuries. Customers of previous ages, even commissioning patrons, did not regard these works as priceless vestiges of The European Genius to be venerated, certainly not as monuments to be protected by State trusts.

On the contrary, they saw them as decorative wall-coverings, or at best mementos of loved ones, to be enjoyed or indeed dismembered Procrustes-like to fit available frames and spaces. And owners about to sell would not hesitate to have "repaired" an existing Master painting in their collections.

Clearly, a wide strip of canvas has been torn or cut from the right-hand edge. We can only speculate about what it depicted: Probably mere generic landscape such as that behind the lady's right shoulder.

As you can see, the template betrays a limited degree of Phidian planning, not slavishly adhered to by the artist. The spiral umbilicus seems to be centered upon Lisa Gherardini's nose, and devolves dextrally to echo the contours of her scalp and her left arm before terminating uncertainly below the triple point of her right arm and the winding road of the background.

But Italian art was never crude, or its formalisms blatant. It never exhibited the commercial mathematicity of the Protestant North, where speed and economy were the premium.

Figure 15.8
A Spiral Galaxy being Golden

And it is to the North that we now turn.

Joseph Mallord Willian Turner was a Londoner and his work and style reflected his craft background. But Turner was also a trained Royal Academician who like all thinking Englishmen worshiped Italy and had in his youth visited the vestiges of Ancient Rome. These inspired some of the light and cheerful watercolors of his early maturity.

Turner was a great innovator who owed little to the past, but in 1829 he re-visited his Homer to paint *Ulysses Deriding Polyphemus*, an oil painting presented on a canvas which is an almost perfect Golden Rectangle.

And that is not by far its only Phidian property.

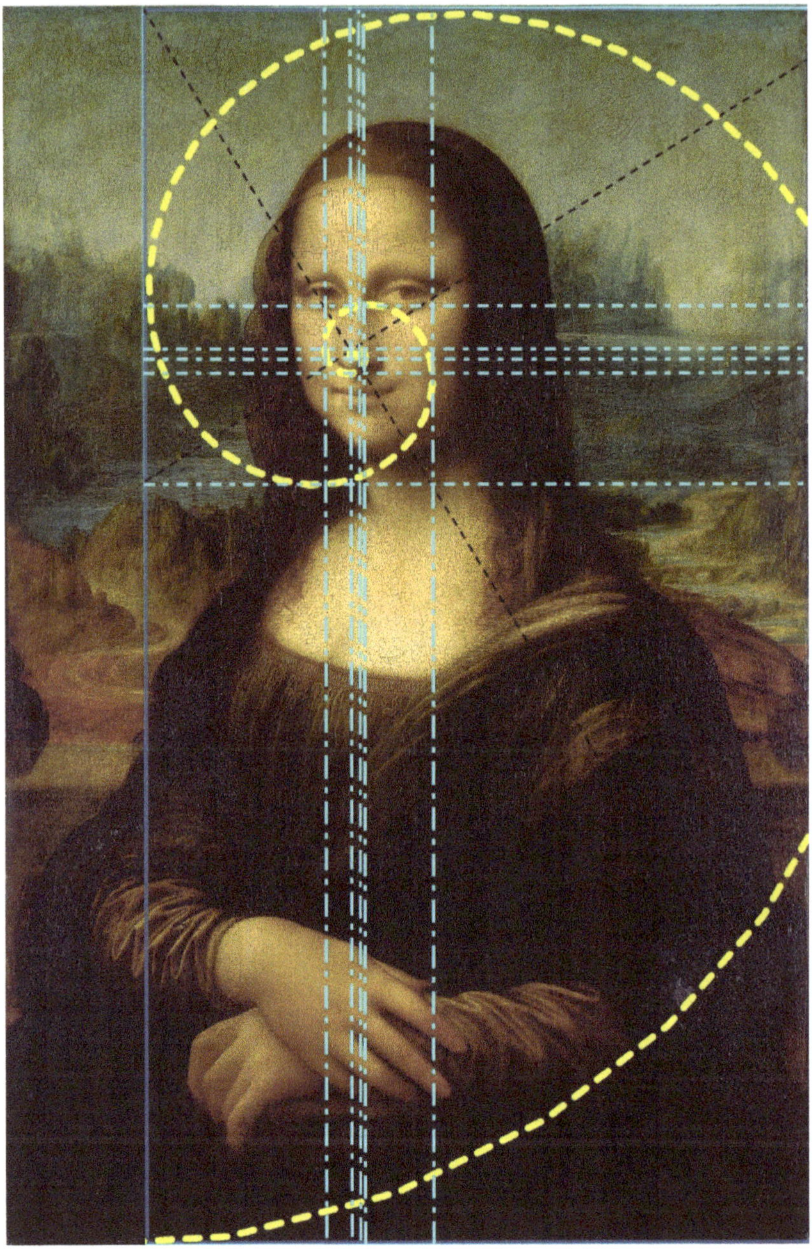

Figure 15.9
The Mona Lisa with a Vincian Phidian
Template Overlay

Figure 15.10 is my rendition of the canvas overlain by a graticular Turnerian template.

In this severely compartmentalised painting, Apollo's golden sunburst is not quite at the umbilicus of the Golden Spiral: It is just to the upper-right of the Φ^4,Φ^4 co-ordinate whilst the top of the mainmast of Ulysses Heroic vessel is at point Φ^6,Φ^6.

Figure 15.10
Ulysses Deriding Polyphemus
by JMW Turner
with Turnerian Template Overlay

For comparison Figure 15.11 is the same canvas with a Vincian overlay. You can now see that the umbilicus of the Golden Spiral is at the top-right of the nearer focal faraglione, whilst the spiral itself seems to echo and guide the contours of the matutinal sail billows on both ships.

Figure 15.11
Ulysses Deriding Polyphemus
by JMW Turner
with Vincian Template Overlay

In his later career, Joseph Turner specialised in seascapes, often stormy, or with a pronounced elegiac tone.

One of the most famous is possibly *The Fighting Temeraire, tugged to her last berth to be broken up, 1838.* In a misty sunset on a flat calm, a steam tug tows the venerable old ship-of-the-line that fought at Trafalgar to the breaker's yard.

I tried a Vincian overlay on this picture because I thought that I had detected an arc described by the ships' mastheads and the top of the tug's funnel. This proved illusory, but a Turnerian template did fit the important elements of the canvas.

When I fitted a Turnerian template it was found that the setting sun was in the center of the Φ^2,Φ^3 block and that the horizontal Φ^4 (i.e. the boundary between the y-axis Φ^2 and Φ^3 block rows) defined the horizon, which matched the waterline of the warship and the bottom of the tug's paddlebox sponsons. The tall, slim funnel of the tug marked the Φ^6 vertical.

The Great Wave off Kanagawa is a colored woodcut print of 1831 by the Japanese artist Katsushika Hokusai. It is Phidian in design. It is not clear to me whether Japan had an aboriginal Phidian aesthetic or whether it was a foreign introduction, possibly by Jesuits.

The spiral umbilicus is clearly at the hydrodynamic still-point of the vast wave whose contours seem to accord with but not ape the Golden Spiral.

Again I am relaxed about the missing right margin for the same reasons that I would not worry about the truncations of an Old Master.

And yet is it a truncation?

A diagonal scribed across the existing print would intersect the summit of Fuji.

Figure 15.12
The Fighting Temeraire, tugged to her last berth
to be broken up, 1838
by JMW Turner
with Turnerian Template Overlay

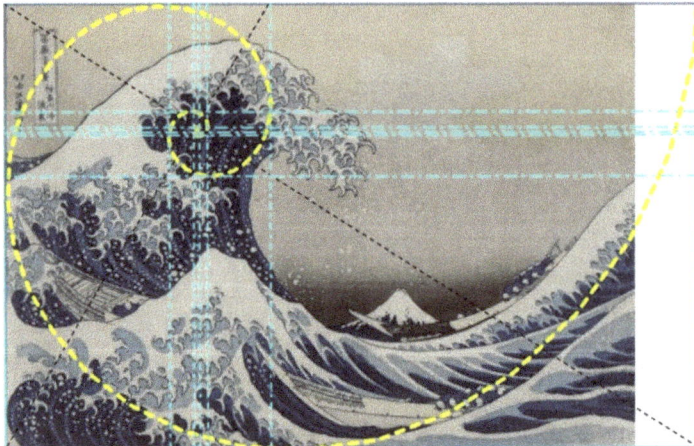

Figure 15.13
The Great Wave off Kanagawa
by Katsushika Hokusai
Using a Vincian overlay

Graphical Exemplars:- B. The Aesthete

The aesthetic tradition is arguably best exemplified by the Pre-Raphaelites and the Impressionists of the last half of the nineteenth-century, and by their twentieth century imitators.

One example, a Suerat, will do for now.

You can see that Suerat has chosen to compose about the black diagonals and the turquoise horizontals and verticals that define the stationary-points ($dy/dx = 0$ or $dy/dx = \infty$). The spiral umbilicus seems to inhabit the lower-right corner of a distant factory, whilst its own or a near horizonal stationary line strongly demarcates a far bridge. A second stationary horizontal follows the base of the second volute and holds the hulls of small pleasure boats whilst also intersecting the nose of the bigger, sitting, boy together with the major diagonal.

(The light aureoles about the bathing boys are *not* artifacts of digital image sharpening: They are present in the original).

Graphical Exemplars:- C. The Arcane Messenger

Arguably the most important of the hermetic painters of sixteenth- and seventeenth-century Europe was Nickolas Poussin (1594-1665). Most or all of his canvases were based upon Biblical or Classical themes spanning a wide diversity of narrative illustrations from Exodus to Virgil to Plutarch, themes and tropes that would have been very familiar to his aristocratic patrons.

Geometry is all over, or should I say under?, the canvases of Nickolas Poussin.

Many libraries could be, and have been, written about Poussin, but we have only space briefly to indicate salient features of his painting *"Et in Arcadia Ego"* which he presented in several versions. The one we will look at is the *Les bergers d'Arcadie* canvas painted around 1638 and now housed at the Musée du Louvre.

The subject is a group of three shepherds and a shepherdess who contemplate a fine tomb in a wilderness. All figures are draped, three in scarlet, saffron, or purple too gorgeous for their profession. They appear to discuss the meaning of a portentous epitaph: "Et in Arcadia Ego".

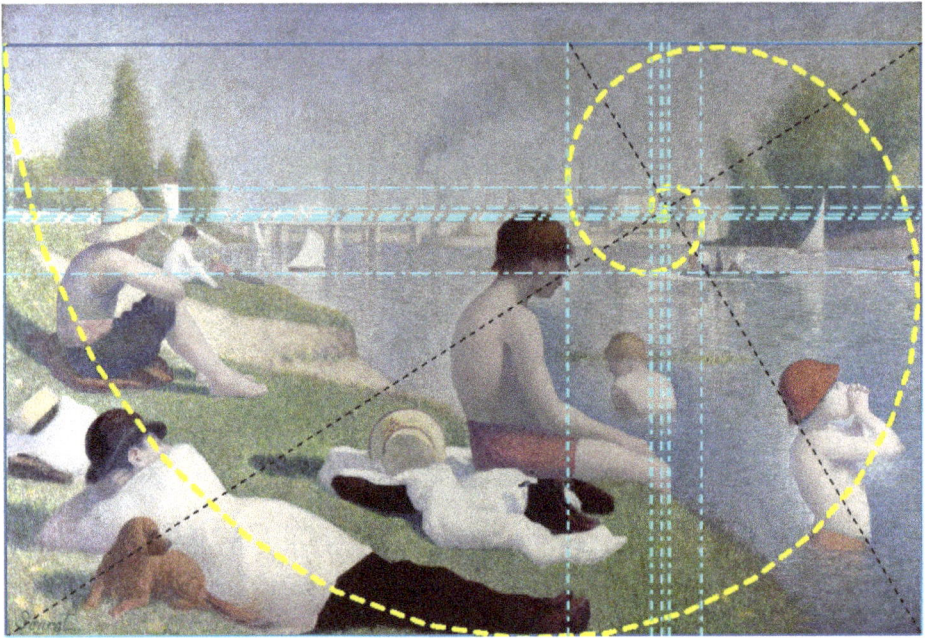

Figure 15.14
Une Baignade, Asnières
by Georges Pierre Seurat,
Using a Vincian overlay

This phrase, from the *Eclogues* of the Roman poet Publius Vergilius Maro, commonly called Virgil, is one of the most obscure and enigmatic ever written in The West. It seems to mean "I am even in Paradise". The Latin is incomplete. Who or what is meant no one knows but the who or what is often assumed, in Virgil's Pagan meaning, to be Satan or Evil. If this is it, then it is a blasphemy to a Christian. The Latin phrase may pre-date the *Eclogues* by a few years as it has been detected in Late Iron Age epigraphy in Britain. There is little evidence that the evil referred to is specifically Death, and the phrase is certainly not a Christian gematria for the obvious reasons already rehearsed.

It is not inconceivable that the phrase *Et in Arcadia Ego* is merely some tribal or collegiate slogan similar to *E Pluribus Unum* or *Nemo Me Impune Lacessit*. It may indeed have been the semi-official motto of The Roman Army. If this is so, then it would not have been the first time a common boast or casual meme had taken on an "almost metaphysical resonance".

Les bergers d'Arcadie is presented in Figure 15.15 without over-printing and in Figure 15.16 with my Vincian template superposed along with two colored regular pentagons that are almost certainly intrinsic to the plan.

Both pentagons are base-upwards and the lower, smaller pentagon is wholly included by the purple upper pentagon. The edge length of the smaller is precisely half the edge length of the larger, implying that the larger, purple pentagon has four times the surface area of the smaller (contrary to illusions): Or, should you elect to sum the two areas, the total area is five times the area of the small, orange pentagon. The larger pentagon's borders just barely include all four human faces, and three of the five visible forearms. The golden spiral umbilicus converges to a gap between the leaves of the background twigs. The gap covers a patch of sky where a face probably appeared, until crudely erased as a pentiment.

The major black diagonal intersects the hearts of the standing shepherdess and her scarlet-clad male companion. Three of the four human faces are within the horizontal, turquoise-colored projections (stationary lines) of the second volute, and the face and upper torse of the female wholly within the second volute.

The total canvas is certainly hermetic but its precise meaning is lost to history. There are many wild theories about the painting and its iconography. It may embody Rosicrucian or Masonic doctrines, either likely to attract capital penalties if published explicitly in the age of Richelieu.

Les Bergers remains in France but many Poussins found their way into aristocratic British dwellings during the course of the seventeenth and eighteenth centuries. I am relaxed about the extensions to the top and bottom of the canvas: These margins may have been requested by a customer or a framer.

Figure 15.15
Et in Arcadia Ego (Les bergers d'Arcadie, 1638)
by Nickolas Poussin

Figure 15.16
Et in Arcadia Ego (Les bergers d'Arcadie, 1638)
by Nickolas Poussin
Using a Vincian overlay

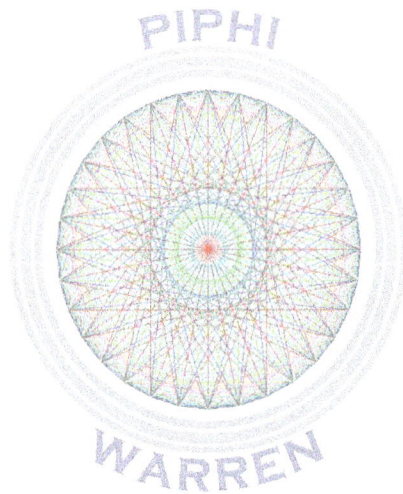

CHAPTER SIXTEEN
PI AND PHI IN BUILDING

Phidian Principles in Architecture

Phidias was an architect and sculptor of Periclean Greece (say 480 – 430 BC). Phidias is said to have participated in the decoration and possibly the design of a splendid hilltop temple called the Parthenon at Athens. The actual architects are thought to have been Iktinos and Callicrates. As aforenoted, these attributions should not be taken literally. I intend no sarcasm when I declare that no documents of patent or priority have survived the passing millennia. I do not know what "Phidias" means. "Iktinos" means something like "kite" or "corvine"; "Callicrates" probably means "The Rule of Beauty" or some cognate abstraction. The temple and its immediate precincts were built between 447 and 438BC, and decoration continued until 432BC, at the acme of Greece.

It is possible that what we call Phidias, Iktinos and Callicrates were all slaves, named and individual only to their friends and masters. This may seem strange to modern sensibilities, schooled to identify Periclean Greece with liberty perfected. The thoughts and attitudes of ancient people differed and to them bondage was not necessarily a badge of shame.

We know from previous observations that Ancient peoples preferred to personalise concepts and abstractions, even up to and including Good and Evil themselves, in a way which modern men despise as childish, indeed dangerous, but which people such as ourselves accept as one way of conceptualising the Inconceivable.

The new Parthenon was, and to an extent is, the votive thank-offering of the people of Athens and the Delian League to their Virgin Goddess Athena for Her vanquishment of the Persian Empire, a defeat to which we owe civilisation in the West. It replaced an older, smaller sanctuary that the Persians destroyed.

Four young virgins inhabited the shelter for a season, not like the aging Vestals of Rome, but girls privileged to weave the purple cloak of the giant golden effigy of the virgin goddess within, and with which the same was enveloped at an annual rite.

There is some evidence that Phidias, or whoever he was, used the Golden Ratio and its derivatives selectively in the planning and drafting of his designs, especially at the Parthenon, and perhaps in his lost sculptures, in addition (controversially) to Parthenon entablature and friezes. The word

"Phidian" has come generally to denote his mathematising school of aesthetics, and there is evidence that his students or imitators in the Classical World used his principles in the planning of sacred precincts, but again much more selectively than some would have you believe.

Too little attention has been given to the practical science, real and imagined, of ancient architects and engineers and in particular the special problems that attend trilithons and suspended beams generally, not only from the point of view of natural seismic shock, common in the Aegean and Asia Minor, but also from the self-weight of structural members in service.[16.1]

The reliable treatments of column-and-lintel static stability is highly mathematical, involving the solution of simultaneous differential equations based upon eighteenth century Euler–Bernoulli Theory and its subsequent sophistications.

This does not mean that Ancient peoples did not have their own ideas or standards regarding reliable engineering and architectural taste. Indeed, many of the temples of ancient Greece and Rome would remain habitable today if it were not for deliberate damage or demolition in modern times.

In general terms we may say that column-and-lintel structural components suffer the following load stresses, all of which, separately or collectively, can result in catastrophic failure:-

Among the relevant stresses are:-

(A)   Beam (Lintel) Stresses
     (i)   Shear Stresses
          Tending to diagonal fractures at the ends
     (ii)   Bending (Flexural) Stresses
          Tending to fan-like cracks along the length, and perpendicular to the length in the beam middle. Failure is usually due to tensile forces at the lower margin.

(B)   Column Stresses
     (i)   Crushing
          Tending to comminute (crush) the body of the support
     (ii)   Buckling
          Tending to make the column snake or curve semi-rigidly:

Ultimately to part near its half-height

This list is far from exhaustive.

Man the Builder

There are four distinct approaches to planning and design:-

(A)   Copernican Science
This is science as commonly understood since the Seventeenth-Century in Europe. It is an attempt to understand the relations within the natural world by means of the systematic application of mathematical methods, broadly defined.

It is probably the most powerful known methodology from the viewpoints of industrial optimisation and reliability, including safety.

Science evolved from the Occamist scholastic doctrine that "accidentals should not be multiplied beyond necessity" and this axiom underpinned the gradual replacement of Ptolemaic cosmology by various heliocentric systems.

In terms of elementary architechtonics, science is exemplified by the Eulerian Load Theory of statics.

(B)   Empirical Methods
These involve measurement and experiment with actual prototypes, at full or reduced scales. Empirical methodologies and observations underpin scientific (A) theory, but (B) methods are complicated by scaling issues and by the typicality or otherwise of materials measured.

In the science of building statics, much of empirical testing and assessment destroys the object.

(C)   Gematric Principles
These depend upon the elaboration of a coded message and its instantiation as metrics of design.

For example, you wish to build a little house for your guinea pig "Jim". Noting that "J" is the tenth letter of the English alphabet; "I" the ninth and "M" the thirteenth letter, you may accordingly fashion him a box thirteen inches wide, ten inches deep and nine high.

There is some circumstantial evidence that the Ancients applied such methods, not least in The Old Testament.

(D)   Phidian Principles

Design principles based upon $\Phi = (1+5^{\frac{1}{2}})/2$ and its corollaries can be applied, and may offer outcomes in optimisation and standardisation, as well as supposed aesthetic benefits.

The Ratios of Phidias have always been controversial and difficult to reconcile with "scientific" parameters, especially π, the Ludolphine Constant.

These difficulties are not merely practical: They reflect the underlying number-theoretical differences between π and Φ.

The implications for constructional safety are as obvious as they are disturbing.

I cannot emphasise too strongly that *none* of these approaches are "correct" in any doctrinal or axiomatic sense. This is despite the fact that each of the four paradigms have assumed an "almost metaphysical resonance" and that all four can be "made to work".

Man strives to create knowing that that only God and Woman can perform the Act.

The first of our Fathers to scrawl a deer on a cave wall knew that his finest effort was a mere imitation, whatever his prayer.

Trilithon Science

The trilithon is the simplest possible static structure you can emplace.

The word "trilithon" comes from the Ancient Greek for "three stones".

It is basically two upright stone members (monolithic or composite) set in or on the ground, and a long beam of rock set atop them, bridging the gap.

Immediately we wonder whether the two uprights can bear the load, even their own weight, and whether the beam will sag in the middle, or break, or whether the set of stones could support any fourth burden.

Stress is a simple mechanical force due to mass and gravity distributed over an area. It is technically a pressure. Any material entity responds to stress with strain, an elastic or plastic change in length or volume, ultimately yielding to breakage. Before breakage, if the stress is removed, elastic deformation strain shall disappear to restore the former shape. On the other hand, plastic deformation is permanent.[16.2,16.3]

An architect would distinguish between a Pier, which is a vertical support member that has a rectangular cross-section normal to applied force; and a Column which has a circular such section.

With regard to suspended Beams, these could equally get called Lintels (especially above a window) or Joist or even, ambiguously, Member.

I will try to clarify the terminology with sketch diagrams:-

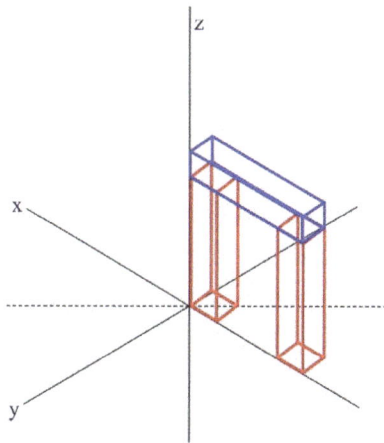

Figure 16.1
A Simple Trilithon

Figure 16.1 represents a basic trilithon with two red piers (square-sectioned columns) bridged by a blue lintel or beam. The piers are 1×1 units square and 5 units tall. The blue beam is of identical dimensions. The columns are entirely covered by the beam ends, so the span is three units, and each column has only to bear half of the weight of the beam without bending (buckling) or crushing. Meanwhile, the beam has to resist the central sag (flexure) due to own weight without splitting and falling between the piers.

Figure 16.2 shows the same simple trilithon loaded with a purple pediment, its Burden. In practice the Burden or Overcroft could be the trilithons share of any supercumbent structure like a stone tile roof, plus the roof's timbers together with further masonry such as a pediment or cross-beams or whatever.

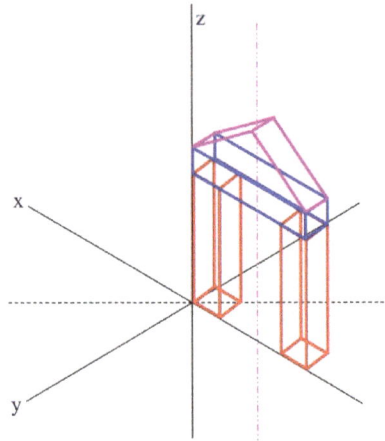

Figure 16.2
A Loaded Trilithon

A sophistication of the loaded trilithon was used by the Ancients to engineer the colonnades of major stone buildings like The Parthenon, a war memorial temple on a hill we call the Acropolis or High City at Athens.

In this design multiple, mutually-reinforcing, trilithons were linked to form load-bearing peristyles. The vertical supports were columns of roughly circular cross-section. The width of these columns was, however, intentionally and systematically varied, whether for aesthetic or tectonic reasons is unknown and remains controversial.

Figure 16.3 illustrates a similar columnar trilithon in isolation:-

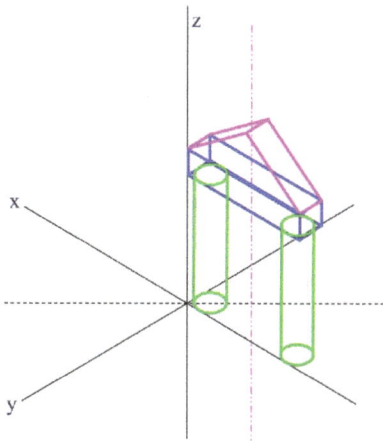

Figure 16.3
A Loaded Columnar Trilithon

It is clear that the beams of these simple structures endure a point load at their centers tending to flex the beam toward breakage. In practice, the beam ends are designed to *overlap* the columns, meeting the next beam or lintel at the *centers* of the column's sections. The tendency of such design is to convert a central point load to a distributed load that is much better supported and permits lighter construction.

In our very basic review of engineering statics we shall explore the implications of trilithon design in terms of Eulerian Stress Theory as pioneered by Leonhard Euler and other mathematical scientists in the first part of the Eighteenth-Century. Pi enters their arguments in interesting ways and we shall see that Phi also comes into play in the design of at least one Classical building, The Parthenon. As always, the relationships are discrete, modest and inconclusive.

By the way, the name Euler is pronounced "oiler" as in the bloke with the greasy rag and the pump-action squirt can.

General Notation

It is probably best to collect all the names and symbols of the physics parameters involved in Euler Statics in one place at the outset of our scientific exposition in order to avoid confusion and the unnecessary tedium of in-text definitions, many of which would be repetitive.

Table 16.1 presents the variable names and symbols for the Beam, and Table 16.2 the names and symbols for the Pier or Column, many of which are the same in concept, and of course in numerical value if the same material is used for both the Piers and the Beam.

			0.618034
Minor Phidian Ratio φ			0.618034
Major Phidian Ratio Φ			1.618034
Ludolphine Constant π			3.141593
Megalithic Cubit ₤			0.454057

			Symbol	Dimensions		
				M	L	T
BEAM						
	Section	Rectangular				
	Substance	Pentelic Marble				
	Input					
		Height	h	0	1	0
		Depth	b	0	1	0
		Span Length	ℓ	0	1	0
		Location of Point Load	y	0	1	0
		Total Length	ℓ+b	0	1	0
		Sectional Area	A_b	0	2	0
		Density	ρ	1	3	0
		Spanning Volume of Beam	V	0	3	0
		Total Volume of Beam	V_b	0	3	0
		Total Mass of Beam	m_b	1	0	0
		Gravitational Acceleration	g	0	1	-2
		Load due to Beam	F_b	1	1	-2
		Width of Burden	w_o	0	1	0
		Height of Burden	h_o	0	1	0
		Depth of Burden	d_o	0	1	0
		Volume of Burden	V_o	0	3	0
		Supported Mass	m_o	1	0	0
		Total Mass At Span	m_t	1	0	0
		Downward Axial Force	P	1	1	-2
		Downward Axial Force per Column	P/2	1	1	-2
		Compressive Strength	τ	1	-1	-2
		Tensile Strength	$σ_b$	1	-1	-2
		Elastic Modulus	E	1	-1	-2
		Shear Wave Velocity	V_s	0	1	1
		Longitudinal Wave Velocity	V_l	0	1	1
	Output	Euler's Critical Load (Zhao)	$σ_z$	1	1	-2
		Failure Force	F_{max}	1	1	-2
		Marginal Ratio (Zhao)	MR_{zhao}	0	0	0
		Mass of Lintel Beam with Burden	M_b	1	0	0
		Top Force on Each Column (F_c)	W_P	1	1	-2
		Break Span (Zhao)	z	0	1	0

Table 16.1
Physics Names and Symbols for
Euler Beam Equations

			Symbol	M	L	T
COLUMN						
	Section	**Circular**				
	Substance	**Pentelic Marble**				
	Input					
		Height	L	0	1	0
		Depth	d	0	1	0
		Width	w	0	1	0
		Mean Radius	r_μ	0	1	0
		Basal Radius	r_d	0	1	0
		Top Radius	r_u	0	1	0
		Column Effective Length Factor	K	0	0	0
		Unsupported Column Length	L	0	1	0
		Effective Length (Tallness)	KL	0	1	0
		Vertical Force on Column	F_c	1	1	-2
		Beam Span Length	ℓ	0	1	0
		Smallest Second Moment of Area	I	0	4	0
		Column Cross Sectional Area	A_c	0	2	0
		Radius of Gyration	R_g	0	1	0
		Slenderness Ratio	λ	0	0	0
		Moment of Inertia	I	0	4	0
	Output	**Euler's Critical Load (F = σ_x)**	σ_z	1	1	-2
		Mass of Column	m_c	1	0	0
		Actual Applied Force	$F_{applied}$	1	1	-2
		Buckling Failure Force	F_{max}	1	1	-2
		Marginal Ratio	MR_{zhao}	0	0	0
		Bessel First Zero	B_{fx}	0	0	0
		Critical Height	h_{crit}	0	1	0
		Compressive Failure Pressure	σ_{ult}	1	-1	-2
		Compressive Failure Force	F_{ult}	1	1	-2
CRITERIA						
	Beam Sag Failure	**Safety Exceeds Unity**	BSF	0	0	0
	Pier Buckle Failure	**Safety Exceeds Unity**	PBF	0	0	0
	Pier Crush Failure	**Safety Exceeds Unity**	PCF	0	0	0
	Beam Sag Failure	**Safety Exceeds Unity**	BSF	0	0	0
	Pier Buckle Failure	**Safety Exceeds Unity**	PBF	0	0	0
	Pier Crush Failure	**Safety Exceeds Unity**	PCF	0	0	0
OTHER RATIOS						
	Euler Loads (Zhao)	**ECL$_{buckle}$/ECL$_{beam}$**	R_{ECL}	0	0	0
	Height to Span Ratio	h/ℓ	R_{hl}	0	0	0
	Test Param 1+2$^{\frac{1}{2}}$	**Q$_1$**	Q_1	0	0	0
	PSD(R_{hl},Q_1)			0	0	0

All Units are SI

Table 16.2
Physics Names and Symbols for
Euler Pier or Column Equations

Euler Beam Theory

Euler Beam Theory also applies to vertical supports, with a few important differences which we may express algebraically.

The Beam

The Euler Critical Stress, dimensionally a pressure, is the load at which a Beam will break, due to downward flexure. Euler Critical Stress is different to the load at which a member might crush, buckle or shear. Euler Critical Stress is essentially a tensile (pulling) load at which a central, lower crack forms and splits the beam.

According to Zhao[16.1], Euler Critical Tensile Stress is given by:-

$$\sigma_x = \frac{3Pl}{b.h^3}.y$$
Equation 16.1

Wikipedia gives a slightly different form:-

$$\sigma_x = \frac{3Pl}{2b.h^3}.y$$
Equation 16.2

That suggests that the Critical Stress is half the Zhao value. Contrary to appearances, the Zhao equation gives a better margin of safety in terms of the beam Span, and therefore I prefer it.

An axial Point Load, P, a *force*, should be imagined at point y along the beam span. We take it that P is concentrated at the beam center and that y is accordingly 0.5

P is given by:-

$$P = g(m_b + m_o)$$
Equation 16.3

As listed $l \equiv \ell$, which is the Span Length of the beam, *excluding* parts of the beam contacting the pier or column supports.

Now clearly some beam materials are stronger than others, and in the context of flexural strength the relevant parameter is the Tensile Strength σ_b, an intrinsic property of the substance, which for Pentelic Marble, we have assumed to be 11500000 Pascals, or 11.5 MPa.

Knowledge of this strength allows us to define the Breaking Force, F_{max}, as:-

$$F_{max} = \frac{2\sigma_b bh^2}{3l}$$

Equation 16.4

Obviously we wish:-

$$\sigma_x < F_{max}$$

Inequality 16.1

For analytical purposes we shall study these relationships, as it were naked, without any consideration of safety margins, or adventitious loading from wind, water, earthquake or whatever.

In responsible design practice we would compute a bearable load on scientific principles and then construct for a bearable load three times bigger ("Brunel's Rule") or six times bigger ("Roebling's Rule"). Proper American or British suspension bridges are designed to tolerate a nose-to-tail phalanx of Abrams or Challenger tanks crossing in a force ten hurricane. For example, during a 105 mile-per-hour wind, the Humber Bridge is designed to bear a load in excess of 38800 tonnes-weight. The central span, the longest in the Western Hemisphere, is 1410 meters long and if 144 Abrams tanks were parked across it (weighing 9641 tonnes altogether) the available margin of safety would be approximately four times.

To return to our simple beam the Beam Sag Failure Criterion, BSF, is given by:-

$$BSF = \frac{F_{max}}{\sigma_x}$$

Equation 16.5

BSF, and similar force ratios, should always exceed unity, else the member will fail at outset.

A Simple Span Failure Model

We would like to know the longest theoretical span that a stiff beam could tolerate before it failed under its own weight. This would implicitly help with safety margins and the comparison of designs. We will call the Failure Span, $z \equiv l_{fail}$.

Because the length 1 is common in both σ_x and F_{max}, and because failure occurs when $\sigma_x = F_{max}$, we can equivalate the two forces and re-arrange the algebra to establish:-

$$z = \sqrt[3]{\frac{\sigma_b b^2 h^5}{2.25P}} = \sqrt[3]{\frac{4\sigma_b b^2 h^5}{9P}}$$

Equation 16.6

z can be surprisingly short, and is usually the historical determinant of building viability.

Euler Column Theory

As noted the cross-section of a pier is square or rectangular, whereas that of a column is circular. The circular section is not necessarily constant, but is usually wider toward the ground.

These differences in sectional geometry and sectional area have implications for load bearing.

We appreciate almost instinctively that a beam or lintel will fail due to sag under its overburden long before it will crumble or buckle. We also sense that it is quite the opposite with tall supports: We expect they shall either buckle or crush long before sagging becomes an issue.

The Effective Height of a pier or column is controlled by the firmness or otherwise with which its top and bottom ends are fixed.

In particular, the science of statics defines a Column Effective Length Factor, K, in these terms:-

K	Column End Conditions
1	Both Ends Pinned: Hinged but Free to Rotate
0.5	Both Ends Fixed
0.699	One End Fixed: The Other End Pinned
2	One End Fixed: The Other End Free to Move Laterally

Table 16.3
Effective Length Criteria

If we define the Actual Column Height as L, then the Tallness or Effective Height, T, is given by:-

$$T = KL$$
Equation 16.7

This enables us to present Euler's Critical Column Load F ≡ σ_z as:-

$$\sigma_z = \frac{\pi^2 EI}{T^2}$$
Equation 16.8

Strictly, σ_z is the Euler Load for Buckling: Crushing (compressive) strength is different.

π is the Ludolphine Constant, the better half of our odd couple, E is the (Young's) Elastic Modulus, dimensionally a pressure, which like Tensile Strength, is a property intrinsic to the material. In the case of common marble E is about 10670000000 Pascals, usually written as 10.67 GPa. Pentelic Marble, a higher-class stone, has an E of about 38480000000 Pascals or 38.48 GPa. E is a measure of elastic restoring-force, or colloquially "stiffness" or nerviness.

I is one of the key parameters of Classical Mechanics. It is the Moment of Inertia, or more technically The Smallest Second Moment of Area. It has the dimensions of length raised to the fourth power (L^4).

For a Pier of rectangular section I is given by:-

$$I = \frac{d.w^3}{12}$$
Equation 16.9

and for a Column of circular section:-

$$I = \frac{\pi . r_\mu{}^4}{4}$$

Equation 16.10

where r_μ is the Mean Column Radius.
The Cross-Sectional Area, A_c, of a Pier is:-

$$A_c = dw$$
Equation 16.11

and for a Column:-

$$A_c = \pi . r_\mu{}^2$$
Equation 16.12

The Radius of Gyration, R_g, is given by:-

$$R_g = \sqrt{\frac{I}{A_c}}$$
Equation 16.13

The Radius of Gyration is dimensionally a length (L^1).
The Slenderness Ratio, λ, is yielded by the expression:-

$$\lambda = \frac{T}{R_g}$$

Equation 16.14

To assess the Load borne by each of the two beams we need to take into account:-

m_b The Mass of the Beam bridging the two Columns

m_o The Mass of the Supercumbent Structures to include for example the share of the roof Mass borne, plus the share of the masonry and other Supported Payload

m_c The Mass of the Actual Pier or Column Itself, usually at the Base

The Mass of the Crossing Beam, M, with Overcroft Paymass is:-

$$M = \rho[bh(l + 2w)] + m_o$$
Equation 16.15

Mass and Weight are distinct, the latter being a force defined by:-

$$F = Mg$$
Equation 16.16

where g is the local Acceleration Due to Gravity, 9.80665 m/s² as a global rough average.

The Column Mass, m_c, may be reconned as:-

$$m_c = \rho dwL$$
Equation 16.17

assuming the Beam and Column are equally deep.

So the applied uniaxial (z-direction) force through each Column, $F_{applied}$, is:-

$$F_{applied} = g \left(m_c + \frac{M}{2} \right)$$

Equation 16.18

The divisor 2 enters because we assume the topload to split equally between the two columns.

Therefore, in context, Euler's Critical Column Load may be expressed:-

$$\sigma_z = \frac{\pi^2 E}{\lambda^2}$$

Equation 16.19

And the Buckling Criterion becomes:-

$$\sigma_{applied} = \frac{F_{applied}}{A_c} < \sigma_z$$

Equation 16.20

Put another way, PBF, the Pier (or Column) Buckle Force is given by:-

$$PBF = \frac{F_{max}}{F_{applied}}$$

Equation 16.21

and should always exceed unity. For common marble forming our simple trilithon it does so by a couple of orders of magnitude.

The Critical Height

A free-standing pier or column possesses a weight of its own, and this weight, bearing down upon its own area could at an ultimate height lead the column to buckle itself like a broken reed.

B_{fx} is a mathematical object known as The First Zero of the Bessel Function of the First Kind: Order -⅓.

The value of B_{fx} is about 1.86635086

The Critical Height, h_{crit}, of a Column is:-

$$h_{crit} = \sqrt[3]{\frac{9B_{fx}^2}{4} \cdot \frac{EI}{\rho g A_c}}$$

Equation 16.22

h_{crit} tends to be of more realistic interest in the context of gracile stand-alone structures such as smokestacks, steeples, pylons and microwave towers (by way of examples).

For a pier made of ordinary marble, having a section one meter square, this height turns out to be around 64 meters, more or less two hundred feet.

I sometimes wonder how the Ancients managed the scale of their more megalomaniacal structures, especially Sulla's Temple of Fortuna Primagenetrix at Palestrina, a structure whose remains I once mistook for the penstocks of a hydroelectric power station from a distance of about eighteen miles.

I also wonder about some of the precarious brick-and-mortar towers of history, knowing that lime mortar creeps plastically over a century or two to form crazy curves in some British mine and mill chimneys that stood long disused into the twenty-first century. Some of the nineteenth century LeBlanc factory chimneys were built up to heights of four or six hundred feet (122 to 183 meters) and stood serviceably for as long as eighty years before orderly demolition.

Modern steel and composite structures are much higher.

Compressive Strength

The Compressive Strength, τ, is another intrinsic physical property of a given substance. It is the uniaxial pressure at which a loaded body will crumble, or in polite language, comminute. In the case of standard-quality marble we may take τ to be 93470000 Pascals, or 93.47 MPa.

Therefore, in the case of a 1×1 meter section pier the applied force could be 93 megapascals before crumble would occur (as opposed to buckle or shear-fracture).

In our example, Pier Crush Failure, PCF, the quotient:-

$$PCF = \frac{F_{crush}}{F_{applied}}$$

Equation 16.23

is about 293, so clearly some other disaster will supervene before our pier crushes under its burden.

Table 16.3 discloses the input and output data for the Beam of our simple marble trilithon, whilst Table 16.4 gives the input and output data for its Piers.

			Symbol	Dimensions M	L	T	Value (SI)
BEAM							
	Section	Rectangular					
	Substance	Marble					
	Input						
		Height	h	0	1	0	1
		Depth	b	0	1	0	1
		Span Length	ℓ	0	1	0	3
		Location of Point Load	y	0	1	0	1.5
		Total Length	ℓ+2b	0	1	0	5
		Sectional Area	A_b	0	2	0	1
		Density	ρ	1	3	0	2690
		Spanning Volume of Beam	V	0	3	0	3
		Total Volume of Beam	V_b	0	3	0	5
		Total Mass of Beam	m_b	1	0	0	13450
		Gravitational Acceleration	g	0	1	-2	9.80665
		Load due to Beam	F_b	1	1	-2	131899.4425
		Supported Mass	m_o	1	0	0	2000
		Total Mass At Span	m_t	1	0	0	15450
		Downward Axial Force	P	1	1	-2	151512.7425
		Downward Axial Force per Column	P/2	1	1	-2	75756.37125
		Compressive Strength	τ	1	-1	-2	93470000
		Tensile Strength	$σ_b$	1	-1	-2	10530000
		Elastic Modulus	E	1	-1	-2	10670000000
		Shear Wave Velocity	V_s	0	1	1	2706.36
		Longitudinal Wave Velocity	V_l	0	1	1	4744.96
	Output	Euler's Critical Load (Zhao)	$σ_z$	1	1	-2	2045422.024
		Failure Force	F_{max}	1	1	-2	2340000
		Marginal Ratio (Zhao)	MR_{zhao}	0	0	0	1.144018189
		Mass of Lintel Beam with Burden	M_b	1	0	0	15450
		Top Force on Each Column (F_c)	W_P	1	1	-2	75756.37125
		Break Span (Zhao)	z	0	1	0	3.137609548

Table 16.3
Input and Output Data for
The Crossing Beam of a
Simple Marble Trilithon

| | | | Symbol | Dimensions | | | |
				M	L	T	Value (SI)
COLUMN							
	Section	**Rectangular**					
	Substance	**Marble**					
	Input						
		Height	L	0	1	0	10
		Depth	d	0	1	0	1
		Width	w	0	1	0	1
		Column Effective Length Factor	K	0	0	0	1
		Unsupported Column Length	L	0	1	0	10
		Effective Length (Tallness)	KL	0	1	0	10
		Vertical Force on Column	F_c	1	1	-2	75756.37125
		Beam Span Length	ℓ	0	1	0	3
		Smallest Second Moment of Area	I	0	4	0	0.083333333
		Column Cross Sectional Area	A_c	0	2	0	1
	Intermediary	**Radius of Gyration**	R_g	0	1	0	0.288675135
		Slenderness Ratio	λ	0	0	0	34.64101615
		Moment of Inertia	I	0	4	0	0.083333333
	Output	**Euler's Critical Load (F = σ_x)**	σ_z	1	1	-2	87757232.47
		Mass of Column	m_c	1	0	0	26900
		Actual Applied Force	$F_{applied}$	1	1	-2	3.395553E+05
		Buckling Failure Force	F_{max}	1	1	-2	8.775723E+07
		Marginal Ratio	MR_{zhao}	0	0	0	258.4475747
		Bessel First Zero	B_{fx}	0	0	0	1.86635086
		Critical Height	h_{crit}	0	1	0	64.16424761
		Compressive Failure Pressure	σ_{ult}	1	-1	-2	9.950000E+07
		Compressive Failure Force	F_{ult}	1	1	-2	9.950000E+07
CRITERIA							
	Beam Sag Failure	**Safety Exceeds Unity**	BSF	0	0	0	1.144018189
	Pier Buckle Failure	**Safety Exceeds Unity**	PBF	0	0	0	258.4475747
	Pier Crush Failure	**Safety Exceeds Unity**	PCF	0	0	0	293.030363
	Beam Sag Failure	**Safety Exceeds Unity**	BSF	0	0	0 OKAY	
	Pier Buckle Failure	**Safety Exceeds Unity**	PBF	0	0	0 OKAY	
	Pier Crush Failure	**Safety Exceeds Unity**	PCF	0	0	0 OKAY	
OTHER RATIOS							
	Euler Loads (Zhao)	ECL_{buckle}/ECL_{beam}	R_{ECL}	0	0	0	42.90421803

All Units are SI

Table 16.4
Input and Output Data for
The Supporting Piers of a
Simple Marble Trilithon

The Central Portico of the Parthenon: The East Range

By reference to Euler Theory and other approximants we shall briefly examine the Eastern part of the outer peristyle of the Parthenon in case it throws any light upon the employment of Pi and Phi by Ancient builders.

Phidias, and his colleagues Iktinos and Callicrates, used a superior marble called Pentelic Marble from quarries a few miles North of Athens.

This is like other marbles a metamorphic limestone, re-crystallised from the sedimentary limestone by stupendous geological forces of temperature, pressure and time. The Ancients treasured marble for use in sculpture and graphic relief work, as well as architecture, because it was flexible enough to stand the impact of chiselling, tough under load, and tolerant of shock. It was also fine-grained and resilient enough to support fine detail and take a high polish, all of which enabled the rock to sustain wear in the pellucid atmospheres of the Classical world: Less so in our age of sulphurous and nitrous air pollution.

The Crushing (Compressive) Strength of Pentelic Marble is 99.5 MPa; the Tensile Strength 11.5 MPa; and the Elastic Modulus 38.48 GPa.

Like virtually all non-metals, Pentelic Marble is a lot more resilient of Crushing than Pulling, in the case of Pentelic Marble nine times as resilient. Therefore, where possible we will spare Pentelic Marble members flexing or bending loads, preferring instead to subject them to pushing forces.

Figure 16.4 shows the East Range of the Parthenon and the red lines are my estimate of where, very approximately, the Ancient stonework would have bridged the columns. The black dotted area, annotated with dimensions in meters, is the approximate area of masonry above the central entablature beam, as it was likely at construction.

Figure 16.4
The East Range of the Parthenon at Athens
Estimation Diagram for the Pedimental Burden of the
Central Two Peristyle Columns

Accordingly, the Volume of this superincumbent stone is $4.3159 \times 5.9795 \times 2.012 = V_b = 51.92353119$ m^3, and the Burden Mass $= 1407.1277$ tonnes. We could discount the stone partly removed to house statuary, but add mass due to part of the roof, etcetera.

These statistics and outcomes are confirmed by Tables 16.5 and 16.6:-

			Symbol	M	L	T	Value (SI)
Minor Phidian Ratio φ							0.618034
Major Phidian Ratio Φ							1.618034
Ludolphine Constant π							3.141593
Megalithic Cubit Ⱡ							0.454057

					Dimensions		
			Symbol	M	L	T	Value (SI)
BEAM							
	Section	Rectangular					
	Substance	Pentelic Marble					
	Input						
		Height	h	0	1	0	1.046
		Depth	b	0	1	0	2.012
		Span Length	ℓ	0	1	0	2.829
		Location of Point Load	y	0	1	0	2.099
		Total Length	$\ell+b$	0	1	0	4.198
		Sectional Area	A_b	0	2	0	2.104552
		Density	ρ	1	3	0	2710
		Spanning Volume of Beam	V	0	3	0	5.953777608
		Total Volume of Beam	V_b	0	3	0	8.834909296
		Total Mass of Beam	m_b	1	0	0	23942.60419
		Gravitational Acceleration	g	0	1	-2	9.80665
		Load due to Beam	F_b	1	1	-2	234796.7394
		Width of Burden	w_o	0	1	0	4.3159
		Height of Burden	h_o	0	1	0	5.9795
		Depth of Burden	d_o	0	1	0	2.012
		Volume of Burden	V_o	0	3	0	51.92353119
		Supported Mass	m_o	1	0	0	140712.7695
		Total Mass At Span	m_t	1	0	0	164655.3737
		Downward Axial Force	P	1	1	-2	1614717.621
		Downward Axial Force per Column	$P/2$	1	1	-2	807358.8103
		Compressive Strength	τ	1	-1	-2	99500000
		Tensile Strength	σ_b	1	-1	-2	11500000
		Elastic Modulus	E	1	-1	-2	38480000000
		Shear Wave Velocity	V_s	0	1	1	0
		Longitudinal Wave Velocity	V_l	0	1	1	0
	Output	Euler's Critical Load (Zhao)	σ_z	1	1	-2	12492236.43
		Failure Force	F_{max}	1	1	-2	5965749.03
		Marginal Ratio (Zhao)	MR_{zhao}	0	0	0	0.477556526
		Mass of Lintel Beam with Burden	M_b	1	0	0	176440.7471
		Top Force on Each Column (F_c)	W_P	1	1	-2	865146.3261
		Break Span (Zhao)	z	0	1	0	2.522189448

Table 16.5
Input and Output Data for
The Crossing Beam of the
East Front Central Columns of
The Parthenon at Athens

		Symbol	M	L	T	Value (SI)
COLUMN						
Section	Circular					
Substance	Pentelic Marble					
Input						
	Height	L	0	1	0	10.4433
	Depth	d	0	1	0	1.7177
	Width	w	0	1	0	1.7177
	Mean Radius	r_μ	0	1	0	0.85885
	Basal Radius	r_d	0	1	0	0.9617
	Top Radius	r_u	0	1	0	0.756
	Column Effective Length Factor	K	0	0	0	1
	Unsupported Column Length	L	0	1	0	10.4433
	Effective Length (Tallness)	KL	0	1	0	10.4433
	Vertical Force on Column	F_c	1	1	-2	906729.2872
	Beam Span Length	ℓ	0	1	0	4.315948
	Smallest Second Moment of Area	I	0	4	0	0.427325846
	Column Cross Sectional Area	A_c	0	2	0	2.317312011
	Radius of Gyration	R_g	0	1	0	0.429425
	Slenderness Ratio	λ	0	0	0	24.31926413
	Moment of Inertia	I	0	4	0	0.427325846
Output	Euler's Critical Load ($F = \sigma_x$)	σ_z	1	1	-2	1488053214
	Mass of Column	m_c	1	0	0	24.20038453
	Actual Applied Force	$F_{applied}$	1	1	-2	9.069666E+05
	Buckling Failure Force	F_{max}	1	1	-2	1.488053E+09
	Marginal Ratio	MR_{zhao}	0	0	0	1640.692385
	Bessel First Zero	B_{fx}	0	0	0	1.86635086
	Critical Height	h_{crit}	0	1	0	127.9075799
	Compressive Failure Pressure	σ_{ult}	1	-1	-2	9.950000E+07
	Compressive Failure Force	F_{ult}	1	1	-2	4.293768E+07
CRITERIA						
Beam Sag Failure	Safety Exceeds Unity	BSF	0	0	0	0.198762208
Pier Buckle Failure	Safety Exceeds Unity	PBF	0	0	0	1640.692385
Pier Crush Failure	Safety Exceeds Unity	PCF	0	0	0	47.34207274
Beam Sag Failure	Safety Exceeds Unity	BSF	0	0	0	FAIL
Pier Buckle Failure	Safety Exceeds Unity	PBF	0	0	0	OKAY
Pier Crush Failure	Safety Exceeds Unity	PCF	0	0	0	OKAY
OTHER RATIOS						
Euler Loads (Zhao)	ECL_{buckle}/ECL_{beam}	R_{ECL}	0	0	0	75.6363482
Height to Span Ratio	h/ℓ	R_{hl}	0	0	0	2.419700145
Test Param $1+2^x$	Q_1	Q_1	0	0	0	2.414213562
PSD(R_{hl},Q_1)			0	0	0	0.226746372

All Units are SI

Table 16.6
Input and Output Data for
The Supporting Columns of the
East Front Central Columns of
The Parthenon at Athens

Table 16.6 clarifies that the entablature monolith fails under its burden. Clearly whatever was above the relevant beam stood for many centuries. So other forces are at play.

I recall many years ago, in a darkened room in Glasgow, studying a very well-directed film with Mr Mark Mathews. The film illustrated the construction of the Parthenon, which Mr Mathews had researched at Athens, and exampled the damage the structure had suffered over the years, not only during war but from the air pollution by vehicles, already a concern at that time.

I was surprised that the Iron Age Greeks had used shaped ties of wrought iron liberally to connect and reinforce the ashlar of the temple. I suppose you will say the clue is in the name, but I was surprised that 2500 years previously they had had the industrial infrastructure and know-how to produce tensile iron on such a scale, and soundly grout it with lead.

This and doubtless other sophistications distributed the load so that parts of the structure too highly stressed were relieved by parts less burdened, so that the total edifice was indefinitely stable.

Another more subtle consideration is that Euler Theory is a *simplification* of physical reality. In particular, P is a Point Load at p = 0.5, the Beam Span Center. The whole fabric of a range of columns is a load *distributor*, and in particular the central Beam suffers distributed load concentrated at its supported ends.

Figure 16.5
The Measured Column Heights
in Relation to Beam Span Length for the
East Front Central Columns of
The Parthenon at Athens

       Firstly, we should clarify that the West Front, largely intact, is likely identical to the East Front in its structural properties. But this author only has positive information about the reduced East Front.

       Note well that in this context, L_{beam}, the Beam Length, is the total length of the beam, not just the free-air span.

       Figure 16.5 illustrates the ratio:-

$$R_{hl} = \frac{h_{column}}{L_{beam}} = \frac{10.4433}{4.3159} = 2.41972705577052$$
Equation 16.24

We may further note that:-

$$Q_1 = 1 + \sqrt{2} = 2.41421356237309$$
Equation 16.25

and:-

$$\frac{\Phi}{Q_1} \approx \frac{2}{3}$$
Equation 16.26

for a PSD:-

$$PSD\left(\frac{\Phi}{Q_1}, \frac{2}{3}\right) = 0.528930823497553$$
Equation 16.27

an error of about one part in 190. My opinion for what it is worth is that Periclean Greeks knew about Φ, π, certainly the Pythagoras's Constant $2^{\frac{1}{2}}$, and the very rough-and-ready approximation of Equation 16.26.

In terms of the Pythagoras's Constant, $2^{\frac{1}{2}}$, a further approximation is:

$$\sqrt{2} \approx \pi - \sqrt{3}$$
Equation 16.28

for:-

$$PSD\left(\sqrt{2}, \pi - \sqrt{3}\right) = 0.330340231240601$$
Equation 16.29

Also:-

$$Q_1 \approx R_{hl}$$
Equation 16.30

for:-

$$PSD(Q_1, R_{hl}) = -0.228376374126934$$
Equation 16.31

or an error of about one part in 438. I do not underestimate the knowledge and precision of the Ancients, but I doubt that the error was appreciated, or worried about if it was.

My opinion is that the builders of the Parthenon intended the central span of the Eastern peristyle to have the Height to Width Ratio $1+2^{\frac{1}{2}}:1$

There is also an interesting high-precision approximation that relates Φ and π in the following way:-

$$\Phi \approx q = \frac{2^{\frac{10}{3}}.5}{33^{\frac{5}{6}}}\left(\frac{5}{\pi} - 1\right) = 1.61795366784591$$

Equation 16.32

for:-

$$PSD(\Phi, q) = 0.004964104867027$$
Equation 16.33

giving an error of one in 20145
Allow the Auxiliary Variables κ and λ as defined below:-

$$\kappa = \frac{-\pi^4 + 10\pi^2 - 1}{\pi^3} = 0.009254673814914$$

Equation 16.34

and:-

$$\lambda = \frac{3\Phi}{2} - 1 - \sqrt{2} = 0.012837420751747$$

Equation 16.35

then:-

$$\frac{\Phi}{1 + \sqrt{2} + \lambda} = \frac{2}{3}$$

Equation 16.36

and:-

$$\frac{1}{\sqrt{2}} \approx A_{approx} = Q_2 = \frac{3}{4}\sqrt{\pi^2\left(1 + \frac{1}{\pi^4}\right) + \kappa.\pi - \frac{82}{9}}$$

$$= 0.707106781186547$$

Equation 16.37

The PSD attaching to Equation 16.37 is given by:-

$$PSD\left(\frac{1}{\sqrt{2}}, Q_2\right) = 0.000000000000079$$

Equation 16.38

This represents an error of about one in $1.26582278481013 \times 10^{15}$.

Substitution for κ and simplification will give the complex exact result:-

$$\frac{1}{\sqrt{2}} = Q_2 = \frac{3}{4}\sqrt{\pi^2\left(1 + \frac{1}{\pi^4}\right) - \left(\pi^2 + \frac{1}{\pi^2}\right) - \frac{8}{9}} = \frac{1}{\sqrt{2}}\sqrt{-1}$$

Equation 16.39

which fixes the relationship between the Pythagoras's Constant and Pi proving that at least that particular transcendental number, π, can be expressed in terms of rational numbers and one irrational number only.

From the above it follows that:-

$$\pi^2 \left(1 + \frac{1}{\pi^4}\right) + \kappa.\pi - 10 = 0$$
$$\textbf{Equation 16.40}$$

and:-

$$\frac{\Phi}{1 + \sqrt{2} + \lambda} - \frac{2}{3} = 0$$
$$\textbf{Equation 16.41}$$

I do not mean to suggest that the Greeks of the Age of Pericles knew these theorems or approximations, or would have cared if they did. What I am inferring is that they could have employed such mathematics given the state of knowledge at the time. You know that they would have managed such information in a geometrical idiom, since algebra is a much later, Arabic construct. Or this is what we think. Classical scholars have estimated that more than ninety-nine percent of the Greek and Roman science documented in 476AD has been lost in the last fifteen hundred years, as war, riot and accident have burned libraries, or damp and mildew decayed organic substrates. Much of what we know of the Classical past has been passed to us by Arabs.

The Parthenon East Range and the Megalithic Cubit

Several have developed concepts of a "Megalithic Cubit" or a "Megalithic Yard" when they have attempted to quantise the metrics implied by surviving structures of the European Iron Age or earlier edifices undocumented.

I suppose they have computed a Highest Common Factor in the spirit of the Millikan and Fletcher Experiment to determine the charge on the electron.

I am grateful to Professor Athanasios G Angelopoulous[16.4] for his precise measurements of the Parthenon at Athens and his quantisation of the dimensions of the building in terms of his Megalithic Cubit, c, (or $\mathbb{C}$) of 0.454056522 meters.

Table 16.7 presents Professor Angelopoulos' findings relative to the key dimensions of the East Front (a peristyle range of eight columns with architrave epistylions, frieze, and surviving pedimental work):-

	Symbol	Cubit Multiplier / Measured Meters	Symbols	Value	Symbolic Value Scaled By Megalithic Cubit	Implied Megalithic Cubit	PSD
Implied Meaglithic Cubit (column height)	c_t	10.4433	-	23	10.4433	0.454056522	
Implied Megalithic Cubit (corner spans)	c_o	9.4075	$2(10\pi/3)$	20.94395	9.50973755		-1.08677
Implied Megalithic Cubit (flanker spans)	c_f	21.5015	$5(3\pi)$	47.12389	21.3969095		0.486434
Implied Meaglithic Cubit (width of Eats Front Third Step)	c_t	30.935		71.20943	32.3331077		-4.5195
Length of Central Architrave Beam (Epistylion)	L_{beam}	4.3159	3π	9.424778	4.2793819		0.846129
Stylobate Third Step Width	W_{tstep}	30.935	$2*(3\pi)+5*(10\pi/3)$	71.20943	32.3331077		-4.5195
Last Two Spans (North: Corner)	V_{nsnsc}	4.722	$10\pi/3$	10.47198	4.75486878		-0.69608
Last Two Spans (North: Flanking)	V_{nsnsf}	4.3	3π	9.424778	4.2793819		0.479491
Facade Height	H			40	18.1622609		

	Symbol	Implied Value	Symbols	Value			PSD
Column Height to Beam Length Ratio	R_{cl}	2.419727056	$1+2^{-X}$	2.414214			0.277856
H/wtstep	R_{hw}	0.561723329	$1/2+\phi/10$	0.561803			-0.01425
Major Ratio of Phidias	Φ	1.61803399					
Minor Ratio of Phidias	ϕ	0.61803399					
Minus Minor Ratio of Phidias	$-\phi$	-0.61803399					

Span from South	Value (meters)	Value (Corners Only)	Value (Flankers Only)
1	4.6855	4.6855	4.3075
2	4.3075		4.293
3	4.293		4.301
4	4.301		4.3
5	4.3		4.3
6	4.3		
7	4.722	4.722	
Sum	30.909	9.4075	21.5015
W_{tstep}	30.935		
PSD	0.0840472		
Mean	4.4155714	4.70375	4.3003
PopSD	0.1825624	0.01825	0.0046

Table 16.7
Parthenon East Front Measurements by
Athanasios G Angelopoulous

Allow that contextually u is the Width of a Flanker Span, and v the Width of a Corner Span. A span is defined as the distance center-to-center of adjacent columns. Clearly, such a distance or any other could be assessed in meters; in Megalithic Cubits; or symbolically, as an implied multiplier of cubits or something.

The Whole Width of the East Range, defined as the Width of the Third Step of the Stylobate, is w_{tstep}, a distance of some 30.935 meters.

Then because the range is of seven columns forming two Corner Spans of v meters each and five Flanker Spans of u meters we may infer:-

$$w_{tseep} = 2v + 5u$$
Equation 16.42

Because:-

$$v = \frac{10}{3}\pi \; cubits = 4.75486877670062 \; meters$$
Equation 16.43

and:-

$$u = 3\pi \; cubits = 9.42477796076938 \; meters$$
Equation 16.44

we are able to make these substitutions:-

$$w_{tstep} \approx 2v + 5u = 2\left(\frac{10}{3}\pi\right) + 5(3\pi)$$
Equation 16.45

or:-

$$w_{tstep} \approx 2v + 5u = \left(\frac{20}{3}\pi\right) + 15\pi = 21.666'.\pi.c \; cubits$$
Equation 16.46

for:-

$$PSD\left(w_{tstep}, \frac{65}{3}\pi.c\right) = 0.091653309991801$$

Equation 16.47

better than an error of one part in a thousand.

Similarly, Professor Angelopoulos established by implication that the Length of the Central Peristyle Architrave Beam (i.e. epistylion) was 3π cubits equivalent to 4.301 meters.

Figure 16.6 shows Professor Angelopoulos's metrics as they relate to the East Front:-

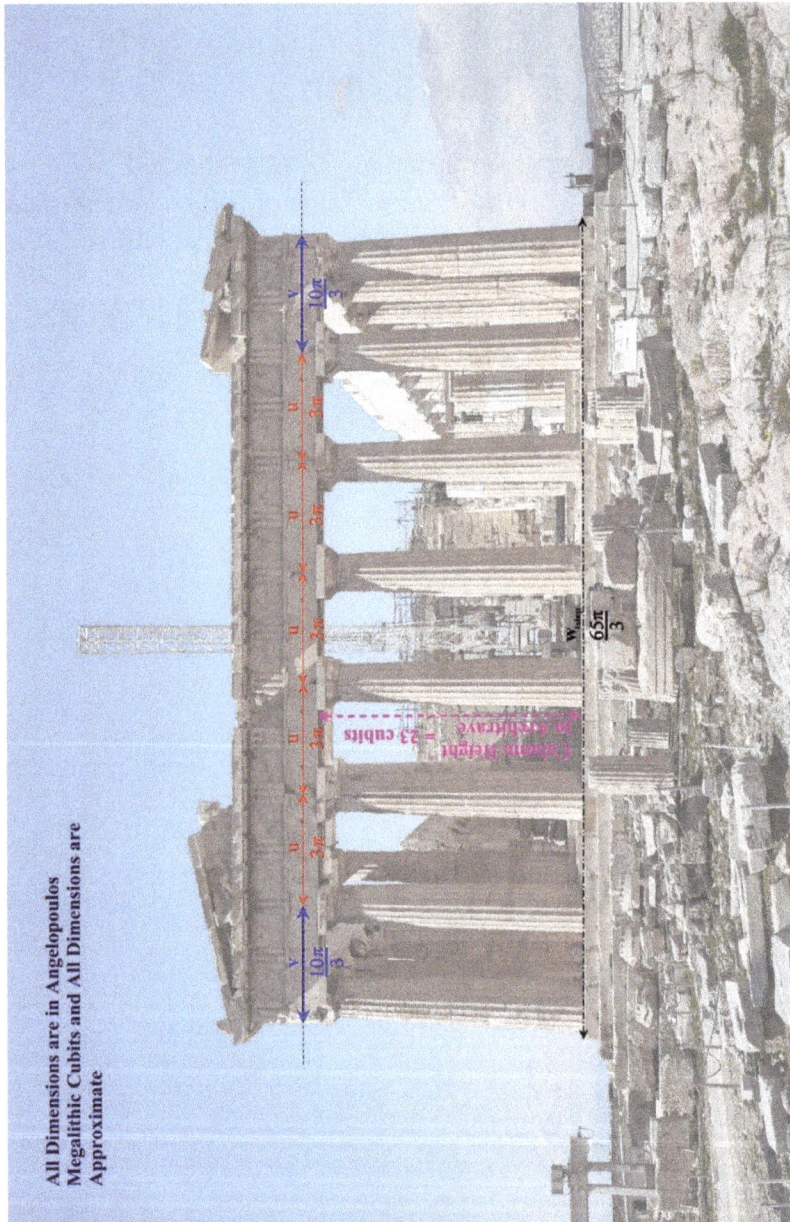

Figure 16.6
Selected Angelopoulos Metrics of the
Parthenon East Front

The East Range and the Golden Rectangle

Fitting Phidian templates to existing or once-existing forms and features is, so to say, a forensic exercise, an art rather than a science. Archaeologists and art historians were as absent as Isaiah when the works were created. Accordingly, they grope in the dark for credible scenarios that, they adjudge, most probably explain the past in the light of current knowledge. But the quest for the Wisdom of the Ancients is a notoriously fraught business, itself the stuff of legend, sometimes of tragedy, sometimes of comedy. I hope to amuse rather than offend.

Like many before me, I have attempted to fit the Golden Rectangle and its included Golden Spiral to one or other of the gable fronts of The Parthenon.

In my opinion the builders assigned no specific role to the spiral umbilicus, but probably used the rectangle and partitions of the rectangle as a *guide* in design.

Figure 16.7 is my interpretation of the Phidian Plan of the East Front.

As you will immediately see, I have assigned Φ to the Width of the Third and Highest Step of the Stylobate (w_{tstep}) but the unitary vertical to an intermediate datum at the foundational surface. That places the ultimate altitude at a point in free air where once the top of a palmate apical acroterion once reached, we may reasonably surmise. (Archaeologists have unearthed fragments of this finial, and re-constructed the ornament).

Given reliable measurements in meters for the height of the central two columns of the East Front and also for the length of their supported epistylion I noticed immediately that this ratio pertained:-

$$R_{hl} = \frac{Column\ Height}{Epistylion\ Length} = \frac{h_{column}}{L_{beam}} = \frac{Q_1}{1} \approx \frac{10.4433}{4.3159}$$
$$\approx 2.41972705577052$$

Equation 16.48

or:-

$$R_{hl} \approx 1 + \sqrt{2}$$
Equation 16.49

for a Percentage Specific Defect of:-

$$PSD(Q_1, R_{hl}) = -0.228376374126934$$
Equation 16.50

an error of about one part in 434
Put more formally if Ψ is the Pythagoras Constant $2^{\frac{1}{2}}$, then:-

$$R_{hl} \approx 1 + \Psi$$
Equation 16.51

In order conveniently to segregate the length parameters addressed below I also offer Figure 16.8

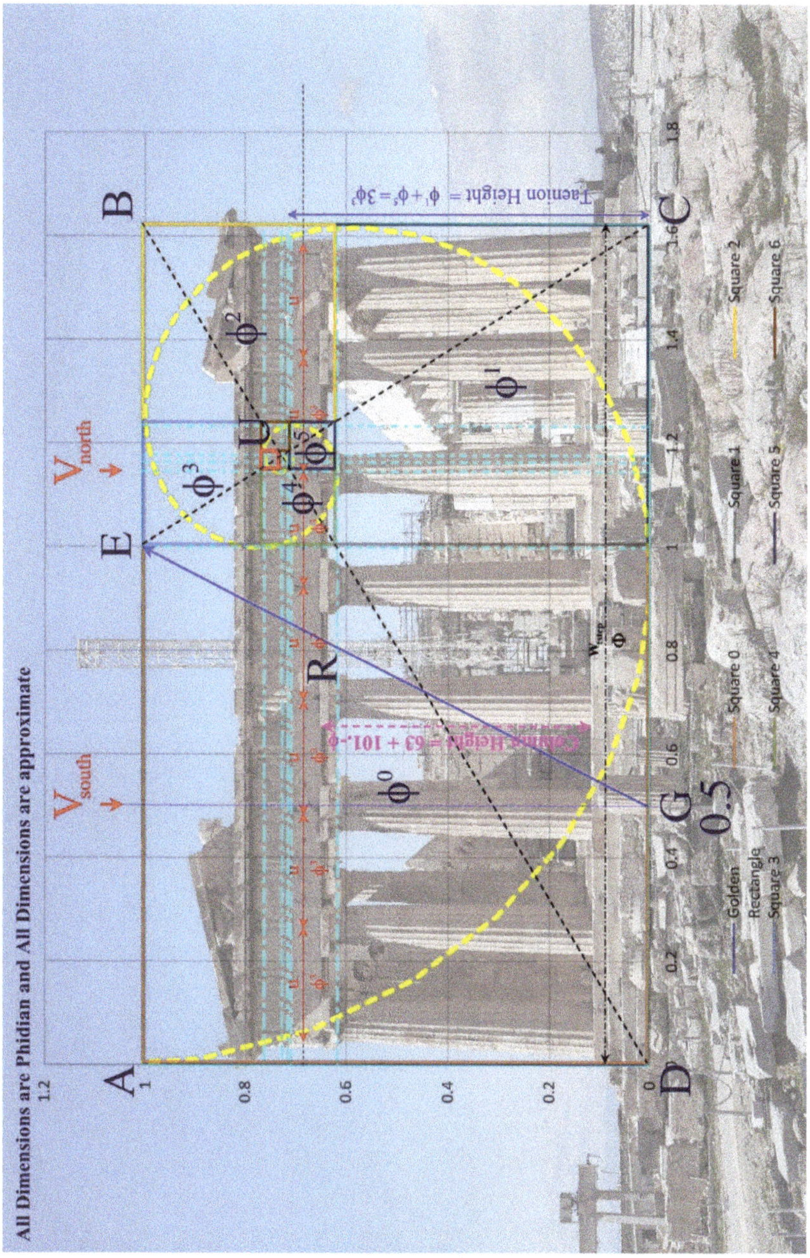

Figure 16.7
The Vincian Template as Fitted to the East Front of
The Parthenon at Athens by
The Author

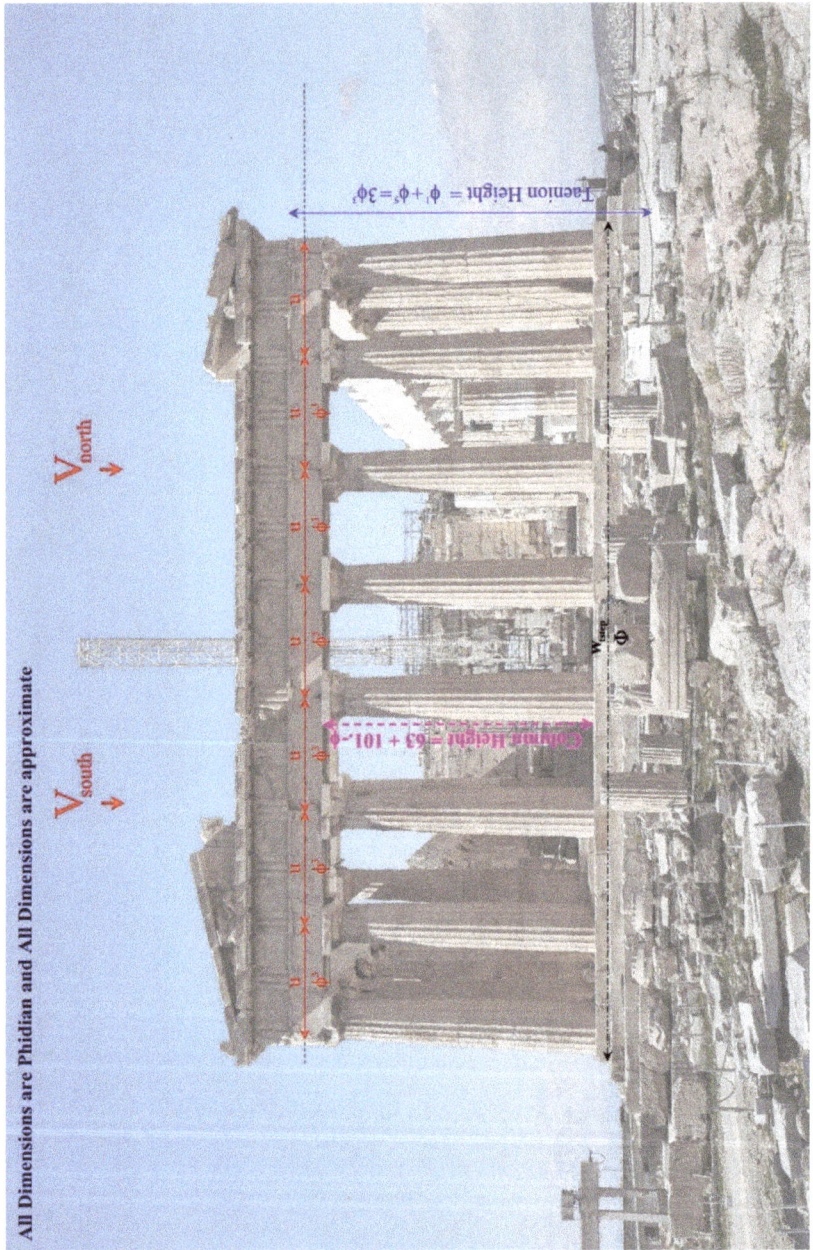

Figure 16.8
Phidian Lengths of the East Front of
The Parthenon at Athens
as Assessed by the Author

Stationary Lines

As aforenoted, the diagram turquoise pecked lines are stationary lines in the calculus sense. The horizonal stationary lines are of zero gradient ($dy/dx = 0$) and the vertical of infinite gradient ($dy/dx = \infty$).

The stationary lines mark left-and-right and top-and-bottom extremes of *particular volutes* of the Golden Spiral. They are accordingly partitions of the Golden Rectangle.

Using my scheme the horizontal stationary line of the second volute intersects the vertical stationary line of the third volute at a point $\phi^0 + \phi^4$ of Φ from the origin of the golden rectangle (i.e. the Point D). This is the medial line of the third column from the North, or in other words the width of the two Northern spans.

Allow that this Width of the two Northern spans together, V_{north}, is the same as that of the two Southern, V_{south}, then:-

$$V_{north} = \Phi - (\phi^0 + \phi^4) = \Phi - (1 + |2 + 3. -\phi|) = 2\phi^3$$
Equation 16.52

V_{north} and V_{south} are numerically equivalent.

Each are two spans of one Corner Span and one Flanker Span.

The implication is that each span of the portico is approximately ϕ^3. Professor Angelopoulos has already pointed out that the end spans are some 11% bigger than the (five) flanker spans. But, with due courtesy, we shall lay this important finding aside whilst we sketch our present scheme.

The Tallness, t, of the South Central Column

The horizontal stationary line that marks the altitude ϕ^1 crosses the capitals of the columns where the echinus joins the abacus. This is not the entire column height.

A superior estimate of t is afforded by:-

$$t = 10\phi^6 + \phi^8 = 10|5 + 8. -\phi| + |13 + 21. -\phi|$$
Equation 16.53

from which:-

$$t = 63 + 101. -\phi$$
Equation 16.54

It is notable that:-

$$\frac{101}{63} = 1.6031746031746 \approx \Phi$$
Equation 16.55

Taenion Height

A taenion is a course of flat marble tiles laid between the architrave and the frieze in a Classical entablature. It was possibly intended to relieve stress by allowing the plastic deformation of the tile and its mortar. This may have protected the sculptural artwork above.
Noting that:-

$$\phi^2 = |F_{n-1} + F_n. -\phi|$$
Equation 16.56

we may reasonably assign the third horizontal stationary line to the course of the taenion. Therefore the level of the taenion is:-

$$h_{tae} = \phi^1 + \phi^5 = 3\phi^3$$
Equation 16.57

As always, there can be differences in both interpretation and the details of interpretation. We are not gods, and we cannot read the minds of the dead.

Whole East Range Portico Width
(approximation of the ideal)

Allow that the Portico Width w_{tstep} is estimated by the Central Span Width due to Column Tallness, u_{column}, such that:-

$$u_{column} = \frac{63 + 101. - \phi}{1 + \sqrt{2}} = 0.23965035458251$$
Equation 16.58

and that:-

$$w_{tstep} = 7u^3$$
Equation 16.59

then:-

$$PSD\left(\frac{\Phi}{7}, \phi^3\right) = -2.12862362522084$$
Equation 16.60

better than one part in 46

$$PSD\left(\frac{\Phi}{7}, u_{column}\right) = 3.67844518356867$$
Equation 16.61

better than one part in 27.
Then the better estimate of Portico Width, w_{tstep}, is:-

$$w_{tstep} = 7\phi^3$$
Equation 16.62

where:-

$$PSD(\Phi, 7\phi^3) = -2.12862362522083$$
Equation 16.63

Therefore our scheme engenders a discrepancy of about 2.1% in the datum length. This is unlikely to be due to builders' error but it may be due to poor modeling on my part, or to earthquake damage or other "land creep".

Pi-Phi Relations as Implicit in the East Front

We can demonstrate that:-

$$LHS = \phi^3.\left(4 + \frac{164 + 101.\Phi}{1 + \sqrt{2}}\right) = 1.00084568451256$$

Equation 16.64

whilst from previous work we know that:-

$$RHS = A_{approx} = \frac{3}{4}\sqrt[3]{\pi^2\left(1 + \frac{1}{\pi^4}\right) - \left(\pi^2 + \frac{1}{\pi^2}\right) + \frac{8}{9}}$$

Equation 16.65

where:-

$$PSD\left(\frac{1}{\sqrt{2}}, RHS\right) = 0$$

Equation 16.66

Given that:-

$$\Psi^{-\frac{1}{2}} = \frac{1}{\sqrt{2}} = LHS = RHS$$

Equation 16.67

we may simplify the equality to yield:-

$$\Psi_{approx} = \frac{1}{164 - 101\Phi}\left(\frac{1.00084568451256}{\phi^3} - 4\right) + 1$$

Equation 16.68

for:-

$$PSD\left(\Psi, \Psi_{appox}\right) = -0.000000000000848$$

Equation 16.69

Also the inverse of the RHS is given by:-

$$RHS = A_{inv} = \frac{4}{3} \cdot \frac{1}{\sqrt{\pi^2\left(1 + \frac{1}{\pi^4}\right) - \left(\pi^2 + \frac{1}{\pi^2}\right) + \frac{8}{9}}}$$
$$= 1.41421356237309$$

Equation 16.70

From which it follows that the Pythagoras Constant, Ψ, is approximated by:-

$$\Psi \approx \frac{4}{3} \cdot \frac{1}{\sqrt{\pi^2\left(1 + \frac{1}{\pi^4}\right) - \left(\pi^2 + \frac{1}{\pi^2}\right) + \frac{8}{9}}}$$

Equation 16.71

The Percentage Specific Defects at this stage are:-

$$PSD\left(\Psi_{appox}, A_{inv}\right) = -0.000000000000864$$
Equation 16.72a
$$PSD\left(\Psi_{appox}, \Psi\right) = -0.000000000000848$$
Equation 16.72b
$$PSD\left(\Psi, A_{inv}\right) = -0.000000000000016$$
Equation 16.72c

Continued simplifications and re-arrangements enable:-

$$\Omega_{approx} = \frac{3}{4}\left(\frac{4 - \dfrac{1.00084568451256}{\phi^3}}{101\phi - 63} + 1\right)$$

Equation 16.73

and:-

$$B_{inv} = \cfrac{1}{\sqrt{\pi^2 \left(1 + \frac{1}{\pi^4}\right) - \left(\pi^2 + \frac{1}{\pi^2}\right) + \frac{8}{9}}}$$
Equation 16.74

Ω_{approx} and B_{inv} are sensibly equivalent, so by squaring both we arrive at the Linkage Equation, $L_{linkage}$:-

$$L_{linkage} = \left[\frac{3}{101\phi - 63}\left(1 - \frac{1.00084568451256}{4\phi^3}\right) + \frac{3}{4}\right]^2$$
$$= \cfrac{1}{\pi^2\left(1 + \frac{1}{\pi^4}\right) - \left(\pi^2 + \frac{1}{\pi^2}\right) + \frac{8}{9}} = \frac{9}{8}$$
Equation 16.75

The relevant Percentage Specific Defects are:-

$$PSD\left(\Omega_{approx}, B_{inv}\right) = 0.000000000000147$$
Equation 16. 76a
$$PSD\left(\Omega_{approx}{}^2, B_{inv}{}^2\right) = 0.000000000000276$$
Equation 16. 76b

Equation 16.75 $L_{linkage}$ appears to me to be an octic algebraic equation but I have not moved forward to an analytic expression at this time. Students who may be interested in this matter may conveniently consult Raghavendra G. Kulkarni[16.5].

A Possible Simplification of $L_{linkage}$

My personal habit is to think of $L_{linkage}$ and things similar as "Non-Occam" mathematical framework objects that are certainly not in their briefest form or commendable as economical industrial utilities.
If they have virtue then it is as open expansions that facilitate the finding and exploration of alternate avenues from the viewpoint of number theoretic science or the applied sciences like building.
Your student could probably identify numerous alternative statements of $L_{linkage}$ that would be more useful. I offer with no recommendation the following essay:-

(a)   Note that:-

$$\phi^3 = \left(\frac{\sqrt{5}-1}{2}\right)^3 = \sqrt{5} - 2$$

Equation 16.77

(b)   Isolate the LHS as Ω_2:-

$$\Omega_2 = \left[\frac{3}{101\phi - 63}\left(1 - \frac{1.00084568451256}{4\{\sqrt{5}-2\}}\right) + \frac{3}{4}\right]^2 = 1.125$$

Equation 16.78

(c)   Simplify LHS:-

$$\Omega_2 = \left(\frac{-0.179737765936883}{101\phi - 63} + \frac{3}{4}\right)^2 = 1.125$$

Equation 16.79

(d)   For brevity, allow that locally:-

$a = 0.0323056645039817$	**Equation 16.80a**
$b = 0.269607$	**Equation 16.80b**
$c = 0.5625$	**Equation 16.80c**
$d = 0.057432292451523$	**Equation 16.80d**
$e = 0.479301333333333$	**Equation 16.80e**
$f = 0.0574323$	**Equation 16.80f**
$g = 0.4793013$	**Equation 16.80g**

(e)   Expand Equation 16.78 by stating its square:-

$$\Omega_2 = a\frac{1}{(101\phi - 63)^2} - b\frac{1}{(101\phi - 63)} + c = 1.12500060683481$$

$$= a\frac{1}{(101\phi - 63)^2} - b\frac{1}{(101\phi - 63)} + \frac{225}{400} = 1.12500060683481$$

$$= a\frac{1}{(101\phi - 63)^2} - b\frac{1}{(101\phi - 63)} + \frac{9}{16} = 1.12500060683481$$

Equation 16.81

(f)   Simplify:-

$$\Omega_2 = \frac{9}{16}\left[d\frac{1}{(101\phi - 63)^2} - e\frac{1}{(101\phi - 63)} + 1\right] = 1.12500060683481$$

Equation 16.82

or to seven decimal places in d and e:-

$$\Omega_2 = \frac{9}{16}\left[f\frac{1}{(101\phi - 63)^2} - g\frac{1}{(101\phi - 63)} + 1\right] = 1.12500058711169$$

Equation 16.83

We are now in the curious position of having a more accurate value for Ω_2 with seven decimal places than with fourteen!

According to Wolfram Alpha®, the Apparent Value of ϕ is 0.626135. Therefore:-

$$PSD(\phi, \phi_{apparent}) = -1.31077115459153$$

Equation 16.84

As a standard quadratic form, Equation 16.83 may be drafted as:-

$$\Omega_2 = f \frac{1}{(101\phi - 63)^2} - g \frac{1}{(101\phi - 63)} - 1 = 0.000001043754112$$

Equation 16.85

We *appear* to have re-defined a putative octic algebraic equation as a much simpler quadratic, but of course the student should be aware that a multitude of detail is implicit in the constants f and g and that notwithstanding Occam we have vitiated a complex scientific system by simplification and superficiality.

Also the LHS of Equation 16.75 for $L_{linkage}$ *credibly* forms an octic (or some algebraic equation) because, as an irrational fraction, Phi is capable in principle of being a root.

On the other hand, the RHS is *incapable* of such a form because Pi is a transcendental number, and by definition cannot be an algebraic root.

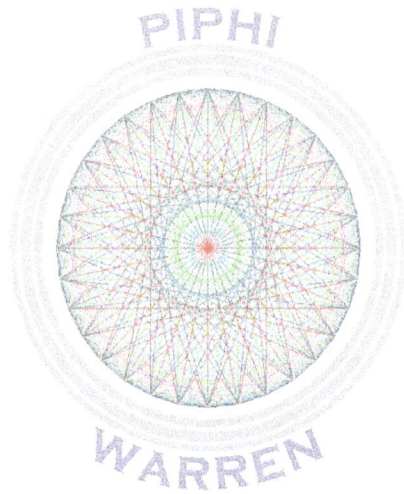

CHAPTER SEVENTEEN
THE OBERG-JOHNSON RELATIONS

The Oberg-Johnson Relations[17.1] are a family of equations of the general conformation:-

$$\pi\Phi = 4(1 + S_B)$$
Equation 17.1

where π is the Ludolphine Constant and Φ is the Major Ratio of Phidias.

S_B is an iterative function of ϕ that converges quickly to an excellent approximation of $\pi\Phi$.

Self-evidently:-

$$\pi = \frac{4}{\Phi}[1 + S_B] = \frac{4}{\Phi}[1 + f(\phi)]$$
Equation 17.2

Oberg and Johnson offer few details of derivation or other theory in their 7 April 2000 paper "The Pi-Phi Product" but very usefully they provide a formulaic EXCEL® spreadsheet to accompany their disquisition, and I deconstructed that to engineer alternative formulations of the Biwabik Sum, $S_B \equiv f(\phi)$. I make no claim for the optimality of my results: Only for their computational effectiveness.

An obvious attraction of The Oberg-Johnson Relations is that they offer a species of direct relationship between π and Φ that is independent of circular functions.

On Page 2 of "The Pi-Phi Product" Oberg and Johnson present:-

$$S_B = 1 + \sum_{k=1}^{\infty} a_k\left[\frac{1}{F_{2k-1} + \Phi F_{2k}}\right]$$
Equation OJ17.1

where $F_1 = 1$; $F_2 = 1$; $F_3 = 2$; and $a_k = b_k + c_k$:-

$$a_k = b_k + c_k$$

Equation OJ17.2

$$b_k = \frac{(-1)^k}{2k+1}$$

Equation OJ17.3

$$c_k = 0$$

Equation OJ17.4

except when k = 3m+1 in which case:-

$$c_{3m+1} = b_m$$

Equation OJ17.5

I was unable to instantiate this system.
Oberg and Johnson also publish:-

$$4 - \int_{k=0}^{\infty} \frac{1}{F_{2k}} = \phi = 1 + \int_{k=2}^{\infty} (-1)^k \cdot \left[\frac{1}{F_k F_{k-1}} \right]$$

Equation OJ17.6

I did not test this because as the authors point out it throws no light upon π.

Oberg and Johnson later present a (partial) explicit series for the product πΦ as:-

$$\pi\Phi = 2^2 \left\{ 1 + \left[\frac{2}{3} \div (F_1 + \Phi F_2) + \frac{1}{5} \div (F_3 + \Phi F_4) - \frac{1}{7} \div (F_5 + \Phi F_6) \right] \right.$$
$$- \left[\frac{2}{9} \div (F_7 + \Phi F_8) + \frac{1}{11} \div (F_9 + \Phi F_{10}) - \frac{1}{13} \div (F_{11} + \Phi F_{12}) \right]$$
$$+ \left[\frac{2}{15} \div (F_{13} + \Phi F_{14}) + \frac{1}{17} \div (F_{15} + \Phi F_{16}) - \frac{1}{19} \div (F_{17} + \Phi F_{18}) \right] \left. \right\}$$
$$= - \cdots$$

Equation OJ17.7

Equation OJ17.7 is very much more complicated than first appearances suggest. For one thing there are implied recurrent series $R_1 = 1,1,1,-1,-1,-1,....$; $R_2 = 1,1,-1,....$; and explicit recurrent series $R_3 = 2,1,1,....$

I was unable to establish generator functions for any of these alternating series.

The Elaborative Spreadsheets

As aforementioned, Oberg and Johnson were considerate enough to supply a spreadsheet that instantiated a process for the computation of their system to yield $\pi\Phi$.

I reproduce the value face of the authors' EXCEL® worksheet in Figure 17.1, as given.

To guide my own experiments I re-constructed this sheet in my own terms. I present that sheet as Figure 17.2. Further to assist my insights I rubricated negative values on my sheet.

The Pi-Phi Product

Formulation by Ed Oberg and Jay A. Johnson developed a formulation for pi-phi, expressing it as as a function of: phi, the number 2, the set of all odd numbers and the set of all Fibonacci numbers, as follows

					Phi	1.61803398874989

a	b	c	d	e	f	a*b*((c/d)/(e+(f*phi))
Sign Term 1	Sign Term 2	Numerator	Odds	Fib N	Fib N+1	Result
1	1	2	3	1	1	0.25464400750007
1	1	1	5	2	3	0.02917960675006
1	(1)	1	7	5	8	(0.00796115571441)
(1)	1	2	9	13	21	(0.00473027472271)
(1)	1	1	11	34	55	(0.00073914715962)
(1)	(1)	1	13	89	144	0.00023889384732
1	1	2	15	233	377	0.00015816550529
1	1	1	17	610	987	0.00002665316787
1	(1)	1	19	1,597	2,584	(0.00000910896167)
(1)	1	2	21	4,181	6,765	(0.00000629590108)
(1)	1	1	23	10,946	17,711	(0.00000109785271)
(1)	(1)	1	25	28,657	46,368	0.00000038579503
1	1	2	27	75,025	121,393	0.00000027288998
1	1	1	29	196,418	317,811	0.00000004852305
1	(1)	1	31	514,229	832,040	(0.00000001733840)
(1)	1	2	33	1,346,269	2,178,309	(0.00000001244261)
(1)	1	1	35	3,524,578	5,702,887	(0.00000000224054)
(1)	(1)	1	37	9,227,465	14,930,352	0.00000000080955
1	1	2	39	24,157,817	39,088,169	0.00000000058673
1	1	1	41	63,245,986	102,334,155	0.00000000010659
1	(1)	1	43	165,580,141	267,914,296	(0.00000000003882)
(1)	1	2	45	433,494,437	701,408,733	(0.00000000002834)
(1)	1	1	47	1,134,903,170	1,836,311,903	(0.00000000000518)
(1)	(1)	1	49	2,971,215,073	4,807,526,976	0.00000000000190
1	1	2	51	7,778,742,049	12,586,269,025	0.00000000000139
1	1	1	53	20,365,011,074	32,951,280,099	0.00000000000026
1	(1)	1	55	53,316,291,173	86,267,571,272	(0.00000000000009)
(1)	1	2	57	139,583,862,445	225,851,433,717	(0.00000000000007)
(1)	1	1	59	365,435,296,162	591,286,729,879	(0.00000000000001)
(1)	(1)	1	61	956,722,026,041	1,548,008,755,920	0.00000000000000
1	1	2	63	2,504,730,781,961	4,052,739,537,881	0.00000000000000
1	1	1	65	6,557,470,319,842	10,610,209,857,723	0.00000000000000
1	(1)	1	67	17,167,680,177,565	27,777,890,035,288	(0.00000000000000)

Sum of above	0.27080092307882
1+Sum	1.27080092307882
2x2 x {1+[Sum]}	5.08320369231526
Pi	3.14159265358979
Phi	1.61803398874989
Pi-Phi	5.08320369231526
Difference	-

Figure 17.1
The Excel 97® Spreadsheet pi-phi.xls
As Provided by Oberg and Johnson

Φ   1.618033988749890
S_b   0.270800923078815
S_a   1.270800923078820
4(1+S_b) 5.083203692315260
πΦ   5.083203692315260

Serial i	a	b	ab	c	Odds 2i+1 d	F_i e	F_{i+1} f	Term of S_B $g=a*b*((c/d)/(e+(f*phii))$	Cumulate Term of S_B	Kernel $g=(1/(2i+1))*(sqrt(5))/(Φ^{2i-1}*(Φ+2)))$	Result $g=ab*c*Kernel$
1	1	1	1	1	3	1	1	0.254644007500070	0.254644007500070	0.127322003735035	0.254644007500070
2	1	1	1	1	5	2	3	0.029179606750063	0.283823614250133	0.029179606750063	0.029179606750063
3	-1	-1	1	-1	7	5	8	-0.007961155714406	0.275862458535727	0.007961155714406	-0.007961155714406
4	-1	1	-1	1	9	13	21	-0.004730274722713	0.271132188813014	0.003651373613356	-0.004730274722713
5	1	-1	-1	1	11	34	55	-0.000739147159617	0.270393036653398	0.000739147159617	-0.000739147159617
6	-1	-1	1	1	13	89	144	0.000238893847319	0.270631930050716	0.000238893847319	0.000238893847319
7	1	1	1	1	15	233	377	0.000158165505286	0.270790090606002	0.000079082752643	0.000158165505286
8	1	1	1	1	17	610	987	0.0000266533167870	0.270816749173872	0.0000266533167870	0.0000266533167870
9	1	-1	-1	1	19	1597	2584	-0.000009108961669	0.270807640212203	0.000009108961669	-0.000009108961669
10	-1	-1	1	-1	21	4181	6765	-0.000006295901081	0.270801344311122	0.000003147950541	-0.000006295901081
11	-1	1	-1	1	23	10946	17711	-0.000001097852711	0.270800246458411	0.000001097852711	-0.000001097852711
12	1	-1	-1	-1	25	28657	46368	0.000000385795027	0.270800632253438	0.000000385795027	0.000000385795027
13	1	1	1	1	27	75025	121393	0.000000272889977	0.270800905143415	0.000000136444989	0.000000272889977
14	1	1	1	1	29	196418	317811	0.000000048523048	0.270800953666463	0.000000048523048	0.000000048523048
15	1	-1	-1	1	31	514229	832040	-0.000000017338403	0.270800936328060	0.000000017338403	-0.000000017338403
16	-1	1	-1	1	33	1346269	2178309	-0.000000012442612	0.270800923885448	0.000000006221306	-0.000000012442612
17	-1	-1	1	-1	35	3524578	5702887	-0.000000002240537	0.270800921644911	0.000000002240537	-0.000000002240537
18	-1	1	-1	-1	37	9227465	14930352	0.000000000809549	0.270800922454460	0.000000000809549	0.000000000809549
19	1	1	1	1	39	24157817	39088169	0.000000000586725	0.270800923041185	0.000000000293363	0.000000000586726
20	1	1	1	1	41	63245986	1.02E+08	0.000000000106589	0.270800923147774	0.000000000106589	0.000000000106589
21	1	1	1	-1	43	1.66E+08	2.68E+08	-0.000000000038820	0.270800923108954	0.000000000038820	-0.000000000038820
22	-1	1	-1	1	45	4.33E+08	7.01E+08	-0.000000000028337	0.270800923080617	0.000000000014169	-0.000000000028337
23	-1	1	-1	1	47	1.13E+09	1.84E+09	-0.000000000005182	0.270800923075435	0.000000000005182	-0.000000000005182
24	-1	1	-1	-1	49	2.97E+09	4.81E+09	0.000000000001898	0.270800923077334	0.000000000001898	0.000000000001898
25	1	1	1	1	51	7.78E+09	1.26E+10	0.000000000001393	0.270800923078727	0.000000000000697	0.000000000001393
26	1	1	1	1	53	2.04E+10	3.3E+10	0.000000000000256	0.270800923078983	0.000000000000256	0.000000000000256
27	1	1	1	-1	55	5.33E+10	8.63E+10	-0.000000000000094	0.270800923078889	0.000000000000094	-0.000000000000094
28	-1	1	-1	1	57	1.4E+11	2.26E+11	-0.000000000000069	0.270800923078819	0.000000000000035	-0.000000000000069
29	-1	1	-1	1	59	3.65E+11	5.91E+11	-0.000000000000013	0.270800923078806	0.000000000000013	-0.000000000000013
30	-1	-1	1	1	61	9.57E+11	1.55E+12	0.000000000000005	0.270800923078811	0.000000000000005	0.000000000000005
31	1	1	1	1	63	2.5E+12	4.05E+12	0.000000000000004	0.270800923078815	0.000000000000002	0.000000000000004
32	1	1	1	1	65	6.56E+12	1.06E+13	0.000000000000001	0.270800923078815	0.000000000000001	0.000000000000001

Figure 17.2
The Pi-Phi Spreadsheet (EXCEL® .xlsx) as Re-Constituted by This Author (Warren)

To establish an orderly and extensible series for the $\pi\Phi$ product I deployed several avenues of attack:-

Control Avenue (1): Binet Expression of Fibonacci Numbers

We have seen elsewhere that a particular Fibonacci Number may directly be accessed using the Binet Formula:-

$$F_k = \frac{\Phi^k - (-\Phi)^{-k}}{\sqrt{5}}$$

Equation 17.3

where k is a Serial Counter (e.g. i), for example $F_7 = 13$
Therefore we are able to write:-

$$F_k + \Phi F_{k+1} = \frac{\Phi^k - (-\Phi)^{-k}}{\sqrt{5}} + \Phi . \frac{\Phi^{k+1} - (-\Phi)^{-(k+1)}}{\sqrt{5}}$$

Equation 17.4

$$F_k + \Phi F_{k+1} = \frac{\Phi^k + \varphi^{k+2}}{\sqrt{5}}$$

Equation 17.5

$$F_k + \Phi F_{k+1} = \frac{\Phi^k (\Phi^2 + 1)}{\sqrt{5}} = \frac{\Phi^k (\Phi + 2)}{\sqrt{5}}$$

Equation 17.6

Control Avenue (2): Simplification of S_B Series Terms

Note that Oberg and Johnson give their spreadsheet equation for $\pi\Phi$ as:-

$$a * b * ((\frac{c}{d})/(e + (f * phi)))$$

Equation OJ17.8

in which phi is Φ.

$a = R_1 = 1,1,1,-1,-1,-1,....$
$b = R_4 = 1,1,-1,1,1,-1,....$
$c = R_3 = 2,1,1,....$
d is the Odd Number $2k+1$
e is F_{2k+1} from $k = 0$
$f = F_{2k+2}$ from $k = 0$
From the forgoing it is clear that the product ab must be:-
$$ab = (1,1),....,-1,-1,-1,1,1,1,-1,-1,-1,1,1,1,....$$
Given these facts it is possible to write:-

$$t_{SB} = ab \left[\frac{\left(\frac{c}{d}\right)}{F_n + \Phi F_{n+1}} \right]$$

Equation 17.7

which using Binet may be re-quoted as:-

$$t_{SB} = ab \left[\frac{\left(\frac{c}{d}\right)}{\frac{\Phi^n.(\Phi + 2)}{\sqrt{5}}} \right]$$

Equation 17.8

or:-

$$t_{SB} = ab \left(\frac{c}{d}\right) . \frac{\sqrt{5}}{\Phi^n.(\Phi + 2)}$$

Equation 17.9

which by substitution may be written:-

$$t_k = ab \left(\frac{2}{2k + 1}\right) . \frac{\sqrt{5}}{\Phi^{2k-1}.(\Phi + 2)}$$

Equation 17.10

whereas aforenoted ab is either 1 or -1.

<u>Control Avenue (3): Segregation of the Two Leading Terms</u>

As we have seen, the first two of the products ab do not conform to the triple alternation -1,-1,-1,1,1,1,....

Accordingly we shall segregate these two leading terms for k = 1 and k =2 and treat them separately.

So for k = 1 :-

$$t_1 = 1 \times 1 \times \left(\frac{2}{3}\right) . \frac{\sqrt{5}}{\Phi^1 . (\Phi + 2)}$$
Equation 17.11a

$$t_2 = 1 \times \frac{1}{2} \times \left(\frac{2}{5}\right) . \frac{\sqrt{5}}{\Phi^3 . (\Phi + 2)}$$
Equation 17.11b

From which we deduce that the Lead Pair, LP, must be:-

$$LP = t_1 + t_2 = \left(\frac{2}{3}\right) . \frac{\sqrt{5}}{\Phi . (\Phi + 2)} + \left(\frac{1}{5}\right) . \frac{\sqrt{5}}{\Phi^3 . (\Phi + 2)}$$
Equation 17.12

or:-

$$LP = \frac{\sqrt{5}}{15\Phi(\Phi + 2)} \left(10 + \frac{3}{\Phi^2}\right)$$
Equation 17.13

which resolves to:-

$$LP = \frac{\sqrt{5}}{15\Phi(\Phi + 2)} \left(10 + \frac{3}{\Phi + 1}\right) = 0.283823614250133$$
Equation 17.14

Control Avenue (4): Resolution of the Negative Series Term Triples

        Once the k =1 and k =2 terms are segregated we may turn our attention to the alternating series of S_B terms proper.
        It is helpful to consider the first triple for k = 3,4,5:-

$$t_3 = -1 \times \frac{1}{2} \times \left(\frac{2}{2k+1}\right) \cdot \left(\frac{\sqrt{5}}{\Phi^{2k-1} . (\Phi+2)}\right)$$

$$= \left(\frac{\sqrt{5}}{\Phi+2}\right) \cdot - \left(\frac{1}{7}\right) \cdot \left(\frac{1}{\Phi^5}\right) = -0.007961155714406$$

Equation 17.15

$$t_4 = -1 \times 1 \times \left(\frac{2}{2k+1}\right) \cdot \left(\frac{\sqrt{5}}{\Phi^{2k-1} . (\Phi+2)}\right)$$

$$= \left(\frac{\sqrt{5}}{\Phi+2}\right) \cdot - \left(\frac{2}{9}\right) \cdot \left(\frac{1}{\Phi^7}\right) = -0.004730274722713$$

Equation 17.16

$$t_5 = -1 \times \frac{1}{2} \times \left(\frac{2}{2k+1}\right) \cdot \left(\frac{\sqrt{5}}{\Phi^{2k-1} . (\Phi+2)}\right)$$

$$= \left(\frac{\sqrt{5}}{\Phi+2}\right) \cdot - \left(\frac{1}{11}\right) \cdot \left(\frac{1}{\Phi^9}\right) = -0.000739147159617$$

Equation 17.17

        You may confirm that these figures tally with both the "Result" terms computed by Oberg and Johnson, and by myself.

Control Avenue (5): Resolution of the Positive Series Term Triples

Once the k =1 and k =2 terms are segregated we may turn our attention to the alternating series of S_B terms proper.

It is helpful to consider the second triple for k = 6,7,8:-

$$t_6 = 1 \times \frac{1}{2} \times \left(\frac{2}{2k+1}\right) \cdot \left(\frac{\sqrt{5}}{\Phi^{2k-1}.(\Phi+2)}\right)$$

$$= \left(\frac{\sqrt{5}}{\Phi+2}\right) \cdot \left(\frac{1}{13}\right) \cdot \left(\frac{1}{\Phi^{11}}\right) = 0.000238893847319$$

Equation 17.18

$$t_7 = 1 \times 1 \times \left(\frac{2}{2k+1}\right) \cdot \left(\frac{\sqrt{5}}{\Phi^{2k-1}.(\Phi+2)}\right)$$

$$= \left(\frac{\sqrt{5}}{\Phi+2}\right) \cdot \left(\frac{2}{15}\right) \cdot \left(\frac{1}{\Phi^{13}}\right) = 0.000158165505286$$

Equation 17.19

$$t_8 = 1 \times \frac{1}{2} \times \left(\frac{2}{2k+1}\right) \cdot \left(\frac{\sqrt{5}}{\Phi^{2k-1}.(\Phi+2)}\right)$$

$$= \left(\frac{\sqrt{5}}{\Phi+2}\right) \cdot \left(\frac{1}{17}\right) \cdot \left(\frac{1}{\Phi^{15}}\right) = 0.00002665316787$$

Equation 17.20

You may confirm that these figures tally with both the "Result" terms computed by Oberg and Johnson, and by myself.

The Consolidation of S_B

We have seen that:-

$$\pi\Phi = 4(1 + S_B)$$
Equation 17.21

and are now in a position to define S_B as a summation of alternating triple-terms of the following form:-

$$S_B = LP + \left(\frac{\sqrt{5}}{\Phi + 2}\right) . T_k = \frac{\sqrt{5}}{15\Phi(\Phi + 2)}\left(10 + \frac{3}{\Phi + 1}\right) + \left(\frac{\sqrt{5}}{\Phi + 2}\right) . T_k$$

Equation 17.22

where the triple-term sum T_k is:-

$$T_k = \sum_{k=1}^{m}(-1)^k . \left[\frac{1}{(6k + 1).\Phi^{6k-1}} + \frac{2}{(6k + 3).\Phi^{6k+1}} + \frac{1}{(6k + 5).\Phi^{6k-3}}\right]$$

Equation 17.23

The counter m gives eight figure accuracy when m = 5 and fifteen-figure accuracy at m = 10. Using Express®, numerical error supervenes beyond m = 10, depending upon the precise configuration of the solution equation.

For good numerical solution the total number of iterations is accordingly near to 3m: Oberg and Johnson used 33 in their worksheet and I used 32 in mine.

Phi in Terms of Pi

It now remains to express π with Φ as the independent variable.

Therefore from Equation 17.2, 17.22 and 17.23:-

$$\pi = \frac{4}{\Phi}\left[1 + \frac{\sqrt{5}}{15\Phi(\Phi + 2)}\left(10 + \frac{3}{\Phi + 1}\right) + \left(\frac{\sqrt{5}}{\Phi + 2}\right) . T_k\right]$$

Equation 17.24

which expands to:-

$$\pi = \frac{4}{\Phi}\left[1 + \frac{\sqrt{5}}{15\Phi(\Phi+2)}\left(10 + \frac{3}{\Phi+1}\right)\right.$$

$$+ \left(\frac{\sqrt{5}}{\Phi+2}\right) \cdot \sum_{k=1}^{m} (-1)^k \cdot \left[\frac{1}{(6k+1).\Phi^{6k-1}}\right.$$

$$\left.\left. + \frac{2}{(6k+3).\Phi^{6k+1}} + \frac{1}{(6k+5).\Phi^{6k-3}}\right]\right]$$

Equation 17.25

The final simplified equation for π that I was able to achieve is Equation 17.26:-

$$\pi = \frac{4}{\Phi}\left\{1 + \phi\left[\frac{1}{3\Phi}\left(2 + \frac{3}{5\Phi^2}\right)\right]\right.$$

$$\left. + \sum_{k=1}^{12} (-1)^k \cdot \Phi^{-6k} \cdot \left(\frac{\Phi}{6k+1} + \frac{2\Phi^{-1}}{6k+3} + \frac{\Phi^{-3}}{6k+5}\right)\right\}.$$

Equation 17.26

The Outcome of Equation 17.26 is 3.14159265358979

The PSD(π,Eqn.17.26) is 0.000000000000014, allowing for numerical error, where π is the value of the MathCad® intrinsic π constant.

I made no attempt to solve either of the implied cubic equations, because I assessed that such would confer no *computational* as opposed to analytic advantage.

The theoreticians tell us that a relationship between π and Φ is impossible because the two numbers are incommensurable, one being transcendental and the other irrational, but as scientists we continue to quest, however hopelessly, for a nice conjugation, using both "nice" and "conjugation" in their Enlightenment senses.

I had hoped, being an optimistic if naive sort of bloke, that we could come up with a pleasant "closed-form" equation similar in spirit to:-

$$Pi = 6\phi^4 + 9\phi^3 + \frac{1}{7}$$

Equation 17.27

But of course life is not like that. The World of the Ancient Greeks and Romans has forsaken us forever, as has that of Tartaglia and Viète.

By the way, for the benefit of our younger students, the value of Equation 17.27 is not π, it is 22/7

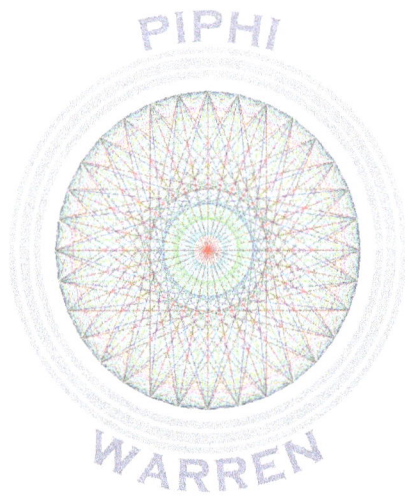

CHAPTER EIGHTEEN
PI , PHI
AND BBP-TYPE FORMULAE

I tried to be a Quaker. Quakers take pride in their rejection of Doctrine. But pride is a sin and the eschewal of doctrine is itself doctrine, rancid and rampant, just as the avoidance of ritual is an observance.

Therefore, we must try to subsist as the other animals who are tentative and vulnerable beneath the tutelage of Someone, awaiting in Ignorance the Delivery of Knowledge.

If we take bread and wine why not? We need not discuss the matter, for talk is interpretation and that incurs judgment. Judgment usurps the office of Someone.

Marriage too is a sacrament.

Marriage is a mystic union. Marriage is compounded of infinite implications, rational and absurd. I strove to wed together Pi and Phi, not as any disgusting importunacy, but as a man of science, one of a perpetual phalanx of intent workers, striving to lay an uncial step toward Understanding.

Marriage is an association, and associations have infinite forms and motives, declarable and inscrutable.

Oberg and Johnson anticipated the simple product that my work has hithertofore only implied.

To be explicit:-

$$T = \pi\Phi$$
Equation 18.1

where T is the (exactly-unknowable) Target Product of Pi and Phi somewhere near the denary irrational number 5.08320369231526. π (or Pi) is the Ludolphine Constant taken to be near 3.14159265358979, and Φ (or Phi) is the Major Ratio of Phidias defined as $(1+5^{1/2})/2$ and which approximates 1.61803398874989

In particular:-

$$\pi\Phi = RHS$$
Equation 18.2

where Right Hand Side (RHS) is, in default of a "closed-form expression", some terminable algebraic process that renders Pi and Phi, so to say, interchangeable components of a rationally-manageable mechanical system.

Some Convenient Functions

Percentage Specific Defect

Percentage Specific Defect is a simple metric of the relative discrepancy of a calculated number from its normative value. It is functionally defined as:-

$$PSD(x, y) = 100 \left(\frac{x - y}{y} \right)$$
Function 18.1

Binomial Coefficient[18.1]

The coefficient of a term of the Binomial Expansion may conveniently be computed in the following manner:-

$$BINOM(n, k) = \binom{n}{k} = \frac{n!}{k! \, (n - k)!}$$
Function 18.2

where n is the Binomial (Expansion) Power and k is the Expansion Term Number. The screech or exclamation mark denotes the single factorial, the product of all whole numbers between 1 and n.

Minus and Plus Bernoulli Numbers[18.2]

In this disquisition where Bernoulli Numbers are referenced without qualification please assume them to be Minus Bernoulli Numbers. Minus Bernoulli Numbers are defined by:-

$$B_-(m) = \sum_{k=0}^{m}\sum_{v=0}^{k}(-1)^v\cdot\binom{k}{v}\cdot\frac{v^m}{k+1}$$
Function 18.3

whilst Plus Bernoulli Numbers are given by:-

$$B_+(m) = \sum_{k=0}^{m}\sum_{v=0}^{k}(-1)^v\cdot\binom{k}{v}\cdot\frac{(v+1)^m}{k+1}$$
Function 18.4

Using fifteen-digit MathCad® Express® on my 64-bit platform I had to use extreme caution with both Binomial Coefficient and Bernoulli Number determinations due to computational errors of rounding and truncation.

Unless you have more powerful resources I recommend that you array high-precision values from a reputable source such as Abramowitz and Stegun which gives ten-figure precision. In any case, it is not really useful to work beyond the first twelve (even) Bernoulli Numbers. Except for $B(1) = -\frac{1}{2}$, all odd Bernoulli Numbers are zero by definition, though I calculated finite tiny values, purely as a result of equipment error!

Table 18.1 presents convenient Bernoulli Numbers.

n	N	D	B_n	B_n As Per A&S Table 23.2	B_n As Per Wikipedia $B_{(m)}$ Formula Using MathCad Express	Defect Table 23.2 versus Wikipedia
0	1	1	1	1	1.00000000000000000	0.00000000000000000
1	-1	2	-0.5	-0.5	-0.50000000000000000	0.00000000000000000
2	1	6	0.166666666666667	0.166666666667000000	0.16666666666666667	0.00000000000033333
3	0	1	0		0.00000000000000000	0.00000000000000000
4	-1	30	-0.033333333333333	-0.033333333333000000	0.03333333333333323	-0.066666666666633323
5	0	1	0		-0.00000000000000114	0.00000000000000114
6	1	42	0.023809523809524	0.023809523810000	0.023809523806591	0.00000000000003409
7	0	1	0		-0.00000000000001592	0.00000000000001592
8	-1	30	-0.033333333333333	-0.033333333333330000	-0.033333333334264171	0.00000000000934171
9	0	1	0		-0.00000000022584572	0.00000000022584572
10	5	66	0.075757575757575758	0.075757575757600000	0.075757781509310	-0.000000205749310
11	0	1	0		-0.000012867152691	0.000012867152691
12	-691	2730	-0.253113553113553	-0.253113553100000	-0.252576947212219	-0.000536605887781
13	0	1	0		-0.009407043457031	0.009407043457031
14	7	6	1.166666666666670	1.166666667000000	1.231109619140630	-0.064442952140630
15	0	1	0		3.414062500000000	-3.441406250000000
16	-3617	510	-7.092156862745100	-7.092156863000000	-3017.906250000000000	3010.814093137000000
17	0	1	0		-276367.75000000000000000	276367.75000000000000000
18	43867	798	54.971177944862200	54.971177940000000	-28086664.00000000000000000	28086718.97117790000000000
19	0	1	0		-524543488.00000000000000000	524543488.00000000000000000
20	-174611	330	-529.124242424242000	-529.124242400000000	-5070168832.00000000000000000	5070168302.87580000000000000

Table 18.1
Bernoulli Numbers

Plouffe Formulae[18.3]

Several Bernoulli-based solutions are Complex.

One of the simple formulae for Pi offered by Simon Plouffe is the following (an un-numbered equation on Page 1 of his 2022 paper):-

$$pi1 \approx \frac{1}{10} \cdot \left(\frac{2n!}{B_n \cdot 2^n} \right)^{\frac{1}{n}} = 3.14159198734778$$

Equation 18.3

For which PSD(π,pi1) = 0.000021207142092

Please note that I have multiplied Plouffe's RHS by 1/10 to make it true.

Equation 18.3 emphasises the point that all estimators of π need to be tested with great circumspection because the different forms differ drastically in their accuracy.

Optimal n is 18, and this yields a real result.

On Page 2 of the said Plouffe paper he further offers:-

$$pi2 \approx \sqrt[n]{\frac{2n!}{B_n \cdot (2^n - 1) \cdot \prod_{j=1}^m \left[1 - \frac{1}{(2j+1)^n} \right]}} = 3.1415926535898$$

Equation 18.4

For which PSD(π,pi2) = -0.000000000000297

In this case the optimal values of integers are m = 4 and n =

14

Bailey-Borwein-Plouffe-type (BBP-type) Formulae[18.4]

The last years of the previous century and the first of this have brought us a radically new approach to the computation of the whole, or of the particular digits, of specified irrational numbers, especially Pi.

BBP-type formulae are generalised as:-

$$S = G \sum_{k=0}^{\infty} \frac{p(k)}{b^{ck} q(k)}$$

Equation 18.5

or in finite terms:-

$$P(n, s, b, m, A) = \sum_{k=0}^{n} \left[\frac{1}{b^k} \sum_{j=1}^{m} \frac{a_j}{(mk + j)^s} \right]$$

Equation 18.6

S is some Number and G is a compound Multiplication Constant, typically a ratio involving rational numbers and possibly irrational numbers. b and c are also Constants that issue from analysis, whilst $p(k)$ and $q(k)$ are both Algebraic Polynomials with integer coefficients ("Diophantines").

The function P is very rapidly convergent so that far from being infinite the sum in Equation 18.6 is "satisfactory" after a mere three to eighteen iterations (if thirteen-digit precision is "good enough").

Accordingly, the ontological integrity of S, as a transcendental number if that is what it is, is never violated.

So n and m are both small Limit Integers whilst k and j are their respective Counters.

s is a Small Exponent, also usually a small integer, almost always unity.

A is a one-dimensional Array of Length m of Small Integers that arise from analysis.

Bailey, Borwein and Plouffe were especially interested in the rapid and accurate approximation of Pi as a whole, or indeed of selected digits along its significand.

By way of example they offered:-

$$pi3 = \sum_{i=0}^{n} \frac{1}{16^i}\left(\frac{4}{8i+1} - \frac{2}{8i+4} - \frac{1}{8i+5} - \frac{1}{8i+6}\right)$$

Equation 18.7

(For my n = 10; PSD(π,pi3) = 0).

The MathCad® Express® value of π at 3.14159265358979 is assumed fiducial.

In terms of the function P(n,s,b,m,A), given that A = (4,0,0,-2,-1,-1), Equation 18.7 may be specified as P(10,1,16,6,a) = 3.14593306268543, for which PSD(π,P(10,1,16,6,a)) = - 0.138159512522286

It is sometimes advantageous to extend the array A as a repeated sequence of integers. For example (Eqn.1.3 Page 3):-

A = (4,0,0,-2,-1,-1,0,0,4,0,0,-2,-1,-1,0,0,4,0,0,-2,-1,-1,0,0)

in this case three periods of eight elements are thus twenty-four, whilst:-

$$pi4 \approx \sum_{i=1}^{24} \frac{a_i}{i \cdot 16^{entier\left(\frac{i}{8}\right)}} = 3.14158739034658$$

Equation 18.8

For PSD(π,pi4) = 0.000167534234761

In a somewhat old-fashioned way I use "entier" to denote the integer part ("floor") of its argument, which in this case is i/8

A numerical improvement on Equation 18.8 is afforded by (Eqn.2.5 Page 4):-

$$pi6 \approx 6\sqrt{\sum_{i=1}^{n} \frac{a_i}{2^i \cdot i^2}} = 3.1415926538847$$

Equation 18.9

where n = 24 and:-

$$A \quad = \quad (1,-3,-2,-3,1,0,1,-3,-2,-3,1,0,1,-3,-2,-3,1,0,1,-3,-2,-3,1,0)$$

PSD(π,pi6) = -0.00000000938723

Integration-based Logarithmic Forms Involving Pi

For convenience sake allow that y = 2.
Then:-

$$\vartheta = 16\left[\frac{1}{8}\ln(2-y^2) - \frac{1}{8}\ln(y^2 - 2y + 2) + \frac{1}{4}\text{atan}\,(1-y)\right]$$
$$= 2ln(2-y^2) - 2ln(y^2 - 2y + 2) + 4\text{atan}\,(1-y)$$
$$= 2\ln(-2) - 2\ln(2) + 4\text{atan}\,(-1)$$
Equation 18.10

Given that y = 2, ϑ is exactly:-

$$\vartheta = -\pi + 2\pi\sqrt{-1}$$
Equation 18.11

and it follows that:-

$$2ln(-2) = 2\left[\ln(2) + \pi\sqrt{-1}\right]$$
Equation 18.12

and:-

$$\pi = -4\text{atan}\,(-1)$$
Equation 18.13

exactly.

With reference to the general function of Equation 18.4, Adegoke (2014) specified that:-

s=1; b=-Φ^6; l=6; j=1...6
whilst I found n = 12 to be optimal.
Further, Adegoke gave:-

A = (Φ^4,0,2Φ^2,0,1,0)

Adegoke's Equation 5.23 was then expressible as:-

$$pi7 = \frac{4}{\Phi^5} . P(n,s,b,l,A) = 3.14159265358979$$
Equation 18.14

For which PSD(π,pi7) = 0.000000000000014
To revert once more to the jargon of the last century I regard this finding as "quantitative" and therefore move forward to observe that:-

$$\Phi^5 = 5\Phi + 3$$
Equation 18.15

whilst re-arrangement of Equation 18.14 gives us:-

$$\frac{\pi\Phi^5}{4} = P(n,s,b,l,A)$$
Equation 18.16

from which by substitution:-

$$\frac{\pi(5\Phi + 3)}{4} = P(n,s,b,l,A)$$
Equation 18.17

Given the Adegoke parameters this was re-expressible as:-

$$T = \pi\Phi = \frac{4}{5}\left[P(12,1,-\Phi^6,6,A) - \frac{3}{4}\pi\right]$$
Equation 18.18

Clearly, sooner or later we shall have to insert an independent estimator for π upon the RHS of Equation 18.18, or effect the equivalent operation by re-arrangement.

Again using the matrix $A = (a_1, a_2, \dots, a_l)$ as:-

$$A = (\Phi^4, 0, 2\Phi^2, 0, 1, 0)$$

we may back-substitute for the function of Equation 18.4 into Equation 18.18 to yield:-

$$T = \pi\Phi = \frac{4}{5}\left[\sum_{k=0}^{n}\left[\frac{1}{b^k}\sum_{j=1}^{m}\frac{a_j}{(lk+j)^s}\right] - \frac{3}{4}\pi\right] = 5.08320369231526$$
Equation 18.19

From Equation 18.19 it obviously follows that:-

$$U = \pi\Phi + \frac{12}{20}\pi = \frac{4}{5}\left\{\sum_{k=0}^{n}\left[\frac{1}{b^k}\sum_{j=1}^{m}\frac{a_j}{(lk+j)^s}\right]\right\} = 6.96815928446914$$
Equation 18.20

or by appropriate substitutions that:-

$$U = \pi(\Phi + 0.6) = \frac{4}{5}\left\{\sum_{k=0}^{n}\left[\frac{(-1)^k}{\Phi^{6k}}\sum_{j=1}^{6}\frac{a_j}{6k+j}\right]\right\} = 6.96815928446914$$
Equation 18.21

At this stage it is also appropriate to recollect that:-

$$-\Phi^6 = -8\Phi - 5 = -17.9442719099992$$
Equation 18.22

By substitution of Equation 18.22 in Equation 18.21 we may produce:-

$$U = \pi(\Phi + 0.6) = \frac{4}{5}\left\{\sum_{k=0}^{n}\left[\frac{(-1)^k}{(-8\Phi - 5)^k}\sum_{j=1}^{6}\frac{a_j}{6k + j}\right]\right\}$$

$$= 6.96815928446914$$

Equation 18.23

Expansion of Equation 18.21 offers us:-

$$U = \frac{4}{5}\left\{\sum_{k=0}^{n}\left[\frac{(-1)^k}{\Phi^{6k}}\sum_{j=1}^{6}\left(\frac{\Phi^4}{6k + 1} + \frac{2\Phi^2}{6k + 3} + \frac{1}{6k + 5}\right)\right]\right\}$$

$$= 6.96815928446914$$

Equation 18.24

At this stage PSD($\pi(\Phi + 0.6)$,U) = 0.000000000000013

As you can see, U is beginning to develop a tripartite structure reminiscent of the Oberg-Johnson system.

Division of the RHS of Equation 18.21 by (Φ+0.6) furnishes pi8, an estimate of π:-

$$pi8 = \frac{4}{5\Phi + 3}\left\{\sum_{k=0}^{n}\left[\frac{(-1)^k}{\Phi^{6k}}\left(\frac{\Phi^4}{6k + 1} + \frac{2\Phi^2}{6k + 3} + \frac{1}{6k + 5}\right)\right]\right\}$$

$$= 3.14159265358979$$

Equation 18.25

PSD(π,pi8) = 0.000000000000014

Restructurings of Equation 18.25 yield:-

$$pi8 = \frac{4}{5\Phi + 3}\left\{\sum_{k=0}^{n}\left[\frac{(-1)^k}{\Phi^{6k}.(6k+1)}\left(\Phi^4 + \frac{(6k+1)2\Phi^2}{6k+3} + \frac{6k+1}{6k+5}\right)\right]\right\}$$
$$= 3.14159265358979$$

Equation 18.26

and:-

$$pi8 = \frac{4}{\Phi^5}\left\{\sum_{k=0}^{n}\left[(-1)^k\left\{\frac{1}{\Phi^{6k-4}.(6k+1)} + \frac{2}{\Phi^{6k-2}.(6k+3)}\right.\right.\right.$$
$$\left.\left.\left. + \frac{1}{\Phi^{6k}.(6k+5)}\right\}\right]\right\} = 3.14159265358979$$

Equation 18.27

and:-

$$pi8 = \sum_{k=0}^{n}\left[(-1)^k\left\{\frac{4}{\Phi^{6k+1}.(6k+1)} + \frac{8}{\Phi^{6k+3}.(6k+3)}\right.\right.$$
$$\left.\left. + \frac{4}{\Phi^{6k+5}.(6k+5)}\right\}\right] = 3.14159265358979$$

Equation 18.28

<u>Pi and the P-function</u>

Recollect that:-

$$P(n,s,b,m,A) = \sum_{k=0}^{n}\left[\frac{1}{b^k}\sum_{j=1}^{m}\frac{a_j}{(mk+j)^s}\right]$$

Equation 18.6

Then:-

$$pi9 = P(1,1,16,6,A) = 3.14565746753247$$
Equation 18.29

where:-

A = (4,0,0,-2,-1,-1)
PSD(π,pi9) = -0.129387046345092

Equation 18.29 is improved by Plouffe's (1995) BBP formula where:-

A = (4,0,0,-2,-1,-1,0,0)

$$pi10 = P(9,1,16,8,A)$$

$$= \sum_{k=0}^{n} \left[\frac{1}{16^k} \left(\frac{4}{8k+1} - \frac{2}{8k+4} - \frac{1}{8k+5} - \frac{1}{8k+6} \right) \right]$$

Equation 18.30

PSD(π,pi10) = 0.000000000000057

Setting n = 4, a reasonable polynomial approximation of pi10 is available as:-

$$pi10^* = \sum_{k=0}^{n} \left[\frac{1}{16^k} \left(\frac{120k^2 + 151k + 47}{512k^4 + 1023k^3 + 712k^2 + 194k + 15} \right) \right]$$

$$= 3.14159600816008$$

Equation 18.31

PSD(p,pi10*) = -0.00010677928848

Adegoke-Layeni Systems[18.6]

Note that:-

$$\sqrt{5} = \sqrt{\frac{5}{3}} \times \sqrt{3} = 2\Phi - 1 = 2.23606797749979$$

Equation 18.32

and the corollary:-

$$\sqrt{3} = \frac{\sqrt{5}}{\sqrt{\frac{5}{3}}} = \sqrt{5} \times \sqrt{\frac{3}{5}} = 1.73205080756888$$

Equation 18.33

Then it follows by arrangement of the Adegoke-Layeni Page 6 (2016) equation that:-

$$\pi_{AL} = \frac{9}{8\sqrt{3}} \times \sum_{j=0}^{n} \frac{(-1)^j}{8^j} \left(\frac{4}{3j+1} + \frac{2}{3j+2} \right) = 3.14159265358979$$

Equation 18.34

where PSD(π,π_{AL}) = -0.000000000000014

Substitutions of the identities of Equation 18.32 and Equation 18.33 enable us to write:-

$$\pi_{AL} = \frac{9}{8\sqrt{\frac{3}{5}}(2\Phi - 1)} \times \sum_{j=0}^{n} \frac{(-1)^j}{8^j} \left(\frac{4}{3j+1} + \frac{2}{3j+2} \right)$$
$$= 3.14159265358979$$
Equation 18.35

for which PSD (π,π_{AL}) = 0; indicating that for fifteen-digit MathCad® Express® the relation is exact.

Simplification of the Integral Multiplier Constant

Note that:-

$$\frac{9}{8\sqrt{\frac{3}{5}}(2\Phi - 1)} = \frac{\frac{9}{8} \times \sqrt{\frac{5}{3}}}{(2\Phi - 1)} = \frac{\left(\frac{3}{\sqrt{8}}\right)^2 \times \sqrt{\frac{5}{3}}}{(2\Phi - 1)} = \frac{3\sqrt{15}}{8} \times \frac{1}{(2\Phi - 1)}$$

$$= 0.649519052838329$$

Equation 18.36

Relevant simplifications enable us to write:-

$$\pi_{AL}(2\Phi - 1) = \frac{3\sqrt{15}}{8} \times \sum_{j=0}^{n} \frac{(-1)^j}{8^j}\left(\frac{4}{3j+1} + \frac{2}{3j+2}\right)$$

$$= 7.02481473104073$$

Equation 18.37

for which PSD $(\pi, \pi_{AL}(2\Phi-1)) = -0.000000000000013$

The Incorporation of BBP Equation 1.2 (pi3) with n = 10

Recollect that:-

$$pi3 = \sum_{i=0}^{n} \frac{1}{16^i}\left(\frac{4}{8i+1} - \frac{2}{8i+4} - \frac{1}{8i+5} - \frac{1}{8i+6}\right)$$

Equation 18.7

for which the computational PSD proved "exact". Then:-

$$T = \pi\Phi = \frac{1}{2}\left[\frac{3\sqrt{15}}{8} \times \sum_{j=0}^{n} \frac{(-1)^j}{8^j}\left(\frac{4}{3j+1} + \frac{2}{3j+2}\right) + \pi\right]$$

$$\approx 5.08320369232852$$

Equation 18.38

But pi3 due to Bailey, Borwein and Plouffe is a "satisfactory" estimate of π and accordingly:-

$$T = \pi\Phi = \frac{1}{2}\left[\frac{3\sqrt{15}}{8} \times \sum_{j=0}^{n} \frac{(-1)^j}{8^j}\left(\frac{4}{3j+1} + \frac{2}{3j+2}\right) + pi3\right]$$

$$\approx 5.08320369232852$$

Equation 18.39

So by substitution of Equation 18.7 in Equation 18.39 we may specify:-

$$T = \pi\Phi = \frac{1}{2}\left[\frac{3\sqrt{15}}{8} \times \sum_{j=0}^{n} \frac{(-1)^j}{8^j}\left(\frac{4}{3j+1} + \frac{2}{3j+2}\right)\right.$$

$$\left. + \sum_{i=0}^{n} \frac{1}{16^i}\left(\frac{4}{8i+1} - \frac{2}{8i+4} - \frac{1}{8i+5} - \frac{1}{8i+6}\right)\right]$$

$$\approx 5.08320369232852$$

Equation 18.40

PSD(T,Eqn.40) is -0.000000000260782. This is eleven-figure accuracy and I ascribe the defect to computational error rather than theoretical shortcomings.

Attempts at the simplification of this long expression using Wolfram Alpha® yielded wrong formulae, and therefore I essayed the following manual simplification:-

$$T = \pi\Phi = \frac{1}{2}[J + K]$$

Equation 18.41

where:-

$$J = \sum_{j=0}^{n} \frac{3\sqrt{15}.(-1)^j}{8^{j+1}} \times \left(\frac{4}{3j+1} + \frac{2}{3j+2} \right) \approx 7.02481473106724$$

$$\approx \pi(2\Phi - 1)$$

Equation 18.42

and:-

$$K = \sum_{j=0}^{n} \frac{1}{16^j} \left(\frac{4}{8j+1} - \frac{2}{8j+4} - \frac{1}{8j+5} - \frac{1}{8j+6} \right)$$

$$= 3.14159265358979 = \pi$$

Equation 18.43

So Equation 18.41 becomes:-

$$T = \pi\Phi = \frac{1}{2} \left[\sum_{j=0}^{n} \left\{ \frac{3\sqrt{15}.(-1)^j}{8^{j+1}} \times \left(\frac{4}{3j+1} + \frac{2}{3j+2} \right) \right. \right.$$

$$\left. \left. + \frac{1}{16^j} \left(\frac{4}{8j+1} - \frac{2}{8j+4} - \frac{1}{8j+5} - \frac{1}{8j+6} \right) \right\} \right]$$

Equation 18.44

This expression is an "exact" identity of T = πΦ according to the PSD yielded by MathCad®.

PIPHI

WARREN

CHAPTER NINETEEN
EPILOG

I had originally intended to call this part of my book the "Conclusion", but then I realised that nothing is concluded, certainly not Scientific Enquiry anyway. Upon consulting Wikipedia for clarification, that excellent resource explained that the Greek-derived word inferred the prospective character of an Epilog in contradistinction to the retrospective nature of a conclusion.

Now the Ludolphine Constant and the Ratio of Phidias, call them what we will, have walked beside us since the dawn of time, and shall continue their stately progress long after we are dead, and as the images of Forms departed may or may not draw traces in the stones like Gryphaea and his cousins.

Whilst it is easy to feel sorry for a couple who are still teaching at their ages, I rather envy Pi and Phi who, staid and stable, retain a courtly regard between unity and a good square four when today the rest of us are at sixes and sevens.

They have been Father and Mother to so much, and so many, including those long-departed cephalopods and cycads who, we may rest Assured, enjoyed their patronage oblivious of any obligation. And in that benign ignorance rest the vast majority of our fellows and contemporaries, unbothered by the pretensions of architecture or the wily obfuscations and betrayals of number theory.

For whilst Our Holy Savior promised that the Truth would set us Free[18.1] it is clear that both Truth and Freedom are unknowable to mortal creatures, and that those must await another State.

We have seen how easy it is to fall into "The Higher Pareidolia" or other delusions, and how the particulars of a Rhombic Triacontahedron (for example) depend upon a fiducial dimension arbitrarily selected by a person, notwithstanding the objective integrity of the ideal geometric solid and its several constant angles.

We have seen that Pi and Phi both have "something to do with" growth, especially the organic growth of individuals, whether those individuals are shellfish or sunflowers or even humans. Arising in part from growth and in part from other concepts we see how Pi and Phi have long appeared the cousins of Time and the uncles of Decay, in our reluctant and ever more halting journey to death, and for this in part have assumed an "almost metaphysical resonance" beguiling of the human imagination.

Pi and Phi are better teachers than were Pythagoras and Plato, because whilst those great magi of Ancient Greece said that Ideals existed and anything that embarrassed the Ideal should be suppressed or ignored, Pi and Phi by their very natures confute the finite and the perfect, for like any truly great conceptions they are both simply stated and profoundly mysterious.

I think you understand that when I accuse "Pythagoras" or "Plato" I intend no libel of dead men. I accuse not the individual but the philosophical school of which he is the nominal, but possibly blameless, progenitor. For the ideal is infinitely corruptible in the will of the collective.

And before people write in, I do understand that English Law does not recognise the libel of the dead, not because justice abhors liars, but because dead men cannot fight duels. Of such are beastly pragma made.

BIBLIOGRAPHY AND REFERENCES

CHAPTER ONE

1.1
"F = phi ≈ 1.618 The Golden Ratio"
Hosted by Gary Meisner from May 15 2012
https://www.goldennumber.net/pi-phi-fibonacci/

1.2
Mathematical Tripos Part II Michaelmas term 2007
Further Complex Methods Dr S.T.C. Siklos
"The arcsin function defined as an integral"
https://www.damtp.cam.ac.uk/user/md327/fcm_2.pdf

CHAPTER TWO

2.1
"Into The Silence"
"The Great War, Mallory and the Conquest of Everest"   2012
Wade Davis
Vintage Books, an imprint of
Random House Publishers of London, UK
Winner of the Samuel Johnson Prize for Non-Fiction 2012
ISBN 978 0 099563 83 9
655pp
Pages 464-467

2.2
"How Archimedes showed that p is approximately equal to 22/7"
Damini D.B.1 and Abhishek Dhar

2.3
Wikipedia contributors. (2022, October 11). Viète's formula. In *Wikipedia, The Free Encyclopedia*. Retrieved 16:31, November 5, 2022, from https://en.wikipedia.org/w/index.php?title=Vi%C3%A8te%27s_formula& oldid=1115360033

2.4
https://keisan.casio.com/exec/system/1354861725

2.5
https://www.intelligence-and-iq.com/g-h-hardy/?utm_source=rss&utm_medium=rss&utm_campaign=g-h-hardy

2.6
"The Curves of Life"
Theodore Andrea Cook
Constable of London   1914
Re-printed by Dover inc. of New York   1979
Paperback
ISBN 0-486-23701-X
479pp excluding rear advertisements
Series of Reciprocals on Page 415

2.7
https://mathworld.wolfram.com/ConvergenceImprovement.html

CHAPTER THREE

3.1
Wikipedia contributors. (2022, October 26). Golden rectangle. In Wikipedia, The Free Encyclopedia. Retrieved 13:15, November 7, 2022, from https://en.wikipedia.org/w/index.php?title=Golden_rectangle&oldid=1118351842

3.2
Wikipedia contributors. (2022, November 28). Ellipse. In Wikipedia, The Free Encyclopedia. Retrieved 15:11, December 3, 2022, from https://en.wikipedia.org/w/index.php?title=Ellipse&oldid=1124354646

CHAPTER FOUR

4.1
"Python Program to Compute the Area and Perimeter of Pentagon"
By Vikram Chiluka
https://python-programs.com/python-program-to-compute-the-area-and-perimeter-of-pentagon/

4.2
"Circular Segment – from Wolfram Mathworld.pdf"
https://mathworld.wolfram.com/CircularSegment.html

4.3
Wolfram Alpha
https://www.wolframalpha.com

4.4
Wikipedia contributors. (2022, October 1). Inverse trigonometric functions. In Wikipedia, The Free Encyclopedia. Retrieved 13:58, October 3, 2022, from
https://en.wikipedia.org/w/index.php?title=Inverse_trigonometric_functions&oldid=1113485991

CHAPTER FIVE

5.1
https://www.wolframalpha.com/input?key=&i2d=true&i=Power%5B%5C%2840%29Divide%5B1%2C2%5D%2BDivide%5Bsqrt%5C%2840%295%5C%2841%29%2C2%5D%5C%2841%29%2Cn%5D

5.2
"The Curves of Life"
Theodore Andrea Cook
Constable of London   1914
Re-printed by Dover inc. of New York   1979
Paperback
ISBN 0-486-23701-X
479pp excluding rear advertisements
Series of Reciprocals on Page 415

CHAPTER SIX

No References

CHAPTER SEVEN

7.1
Wikipedia contributors. (2022, October 1). Inverse trigonometric
functions. In Wikipedia, The Free Encyclopedia. Retrieved 13:58, October
3, 2022, from
https://en.wikipedia.org/w/index.php?title=Inverse_trigonometric_function
s&oldid=1113485991

7.2
"An Innovative Method for Approximating Arcsine Function"
Slim Ben Othman and Yogesh J Bagul
www.preprint.org   Posted 26 July 2022
doi:10.20944/preprints202207.0388.v1
Mathematics Subject Classification (2010): 26E99, 33B10, 42A10
Keywords: arcsine function; approximation; interpolation; Shafer-Fink
inequality; pade approximants

7.3
"A generalization of the Shafer-Fink inequality"
Jacopo D'Aurizio
University of Pisa
arXiv:1304.0753v1
https://arxiv.org/abs/1304.0753v1
10 March 2013
9pp
Theorem 5

CHAPTER EIGHT

8.1
"Handbook of Mathematical Functions with Formulas, Graphs and Mathematical Tables"
Edited by Milton Abramowitz and Irene A Stegun
National Bureau of Standards Applied Mathematics Series * 55
United States Department of Commerce
Issued June 1964
Tenth Printing, December 1972 with Corrections
1046pp

8.2
https://mathworld.wolfram.com/Newton-CotesFormulas.html

8.3
Wikipedia contributors. (2022, October 22). Simpson's rule. In Wikipedia, The Free Encyclopedia. Retrieved 11:30, October 28, 2022, from https://en.wikipedia.org/w/index.php?title=Simpson%27s_rule&oldid=1117538854
MLA Style Manual

CHAPTER NINE

9.1
https://www.wolframalpha.com/input?i=cos%28x%29*acos%28y%29

CHAPTER TEN

10.1
Benammar Et Al

CHAPTER ELEVEN

11.1
"The Divine Proportion: A Study in Mathematical Beauty"
Herbert Edwin Huntley   1970
Dover Publications of New York and
Constable of London

ISBN 0-486-22254-3
186pp excluding advertisements
Analysis of Beauty p.45
EXERCISE III - CONE PROBLEM
"The circumference of the base of a right circular cone, of which
the semi-vertical angle is 54 degrees and the slant side measures one foot,
is pi.phi feet, and the curved surface is 0.5pi.phi square feet.
The proof of this, with the help of the table on p.40, is left to the reader."

11.2
Wikipedia contributors. (2022, December 15).
Cone.
In Wikipedia, The Free Encyclopedia.
Retrieved 11:28, December 20, 2022, from
https://en.wikipedia.org/w/index.php?title=Cone&oldid=1127602296

11.3
https://www.wolframalpha.com/input?i2d=true&i=atan%5C%2840%29Di
vide%5B%5C%2840%29Divide%5Bphi%2C2%5D%5C%2841%29%2Cs
qrt%5C%2840%29291-
Power%5B%5C%2840%29Divide%5Bphi%2C2%5D%5C%2841%29%2
C2%5D%5C%2841%29%5D%5C%2841%29

11.4
"Exact Coordinates of Rhombic Solids"
Izidor Hafner
November 2012
CC BY-NC-SA
https://demonstrations.wolfram.com/ExactCoordinatesOfGoldenRhombic
Solids/
Triacontahedron

11.5
"Rhombic Triacontahedron"
ENCYCLOPEDIA POLYHEDRA
RHOMBIC TRIACONTAHEDRON
Robert W Gray
August 5 2007

http://www.rwgrayprojects.com/rbfnotes/polyhed/PolyhedraData/Rhombic Triaconta/RhombicTriaconta.pdf

11.6
"Rhombic triacontahedron"
Various Contributors
Polytope Wiki
https://polytope.miraheze.org/wiki/Rhombic_triacontahedron

11.7
Wikipedia contributors. (2022, October 21).
Rhombic triacontahedron.
In Wikipedia, The Free Encyclopedia.
Retrieved 16:29, January 4, 2023, from
https://en.wikipedia.org/w/index.php?title=Rhombic_triacontahedron&old id=1117458093

11.8
"Rhombic triacontahedron"
Various Contributors
Polytope Wiki
https://polytope.miraheze.org/wiki/Rhombic_triacontahedron

11.9
"A Geometric Analysis of the Platonic Solids and other Semi-Regular Polyhedra:
With an Introduction to the Phi Ratio"
Kenneth James Michael MacLean
The Loving Healing Press of Ann Arbor   2007
ISBN 978-1-932690-99-6   (Hardback)
ISBN 978-1-932690-99-9   (Softback)
153pp
USD 34.95   GBP 24.95
Diagrams drawn using WinGeom by Rick Parris
Now at https://www.softpedia.com/get/Science-CAD/Wingeom.shtml
        (Jan 2023)
Chapter 14: p122

11.10
"Rhombic triacontahedron"
Various Contributors
Polytope Wiki
https://polytope.miraheze.org/wiki/Rhombic_triacontahedron

11.11
"Rhombic Triacontahedron"
ENCYCLOPEDIA POLYHEDRA
RHOMBIC TRIACONTAHEDRON
Robert W Gray
August 5 2007
http://www.rwgrayprojects.com/rbfnotes/polyhed/PolyhedraData/Rhombic
Triaconta/RhombicTriaconta.pdf

11.12
Wikipedia contributors. (2022, October 21).
Rhombic triacontahedron.
In Wikipedia, The Free Encyclopedia. Retrieved 11:12, March 15, 2023, from
https://en.wikipedia.org/w/index.php?title=Rhombic_triacontahedron&old
id=1117458093

11.13
"A Geometric Analysis of the Platonic Solids and other Semi-Regular Polyhedra:
With an Introduction to the Pi Ratio"
Kenneth James Michael MacLean
The Loving Healing Press of Ann Arbor   2007
ISBN 978-1-932690-99-6   (Hardback)
ISBN 978-1-932690-99-9   (Softback)
154pp
Diagrams drawn using WinGeom by Rick Parris
Now (Jan 2023) at:-
https://www.softpedia.com/get/Science-CAD/Wingeom.shtml
Chapter 14:p122

CHAPTER TWELVE

12.1
Genesis 19:37-38

12.2
Wikipedia contributors. (2022, September 5). Blue Lias. In Wikipedia, The Free Encyclopedia. Retrieved 12:51, November 23, 2022, from https://en.wikipedia.org/w/index.php?title=Blue_Lias&oldid=1108689241

12.3
Wikipedia contributors. (2022, August 24). Gryphaea. In Wikipedia, The Free Encyclopedia. Retrieved 12:52, November 23, 2022, from https://en.wikipedia.org/w/index.php?title=Gryphaea&oldid=1106417761

12.4
File:Gryphaea arcuata fossil oyster (Blue Lias, Lower Jurassic; coastal cliffs near Lyme Regis, far-western Dorset County, southwestern England) 1 (15206675956).jpg
James St. John, CC BY 2.0 <https://creativecommons.org/licenses/by/2.0>, via Wikimedia Commons

12.5
Isaiah 55:8

12.6
http://www.ams.org/publicoutreach/feature-column/fcarc-shell5

12.7
http://www.ams.org/publicoutreach/feature-column/fcarc-shell3

12.8
"Growth Stages in Some Jurassic Ammonites"
Ethel D Currie, BSc, PhD, FGS
Transactions of the Royal Society of Edinburgh
Vol. LXI. Part I. (No.6)
Published by Oliver and Boyd of Edinburgh and London
1944
pp 171-199

https://csef.usc.edu/History/2009/Projects/J1606.pdf

13.2
Wikipedia contributors. (2022, November 22).
Parastichy.
In *Wikipedia, The Free Encyclopedia*
Retrieved 14:41, January 18, 2023, from
https://en.wikipedia.org/w/index.php?title=Parastichy&oldid=1123236805

13.3
"The Divine Proportion: A Study of Mathematical Beauty"
Herbert E Huntley
Dover Books on Mathematics   1 September 1970
ISBN-13 978-0486222547
185pp (paperback)
Page 163 Fig.12.5

CHAPTER FOURTEEN

14.1
"The Discovery of Quasicrystals"
Sven Liden
Kungl. Vetenskaps Akademian
The Royal Swedish Academy of Sciences
Professor of Inorganic Materials Chemistry
Lund University
Member of the Nobel Committee for Chemistry
5 Oktober 2011
https://www.nobelprize.org/uploads/2018/06/advanced-
chemistryprize2011.pdf

14.2
Locality: Shangbao Pyrite mine, Leiyang County, Hengyang Prefecture,
Hunan Province, China (Locality at mindat.org)
Size: 3.4 x 3.4 x 3.4 cm.
Bright, golden, beautifully striated pyrite crystals.
Photo: Robert M Lavinsky
Atribution: Rob Lavinsky, iRocks.com – CC-BY-SA-3.0
https://commons.wikimedia.org/wiki/File:Pyrite-184681.jpg

14.3
File:Pyrite-193871.jpg
https://en.wikipedia.org/wiki/File:Pyrite-193871.jpg
Photo: Robert M Lavinsky:
Source: http://www.mindat.org/photo-193871.html
Attribution: Rob Lavinsky, iRocks.com – CC-BY-SA-3.0
Locality: Magma Mine (Magma Superior Mine; Irene claim; Hub claim; Pomeroy; Superior Division; Silver Queen; Monarch claim; Magma Copper Mine; Broken Hill; Apex), Superior, Pioneer District, Pinal Mts, Pinal County, Arizona, USA (Locality at mindat.org)
Size: 3.2 x 3.2 x 2.6 cm.
A textbook, beautiful, complete all-around pyritohedron from the Magma Mine and the Richard Gaines Collection. This superb, pristine crystal has mirror-bright, brass-yellow luster and all of the faces have exquisite striations.

14.4
File: Pyrite-193871 angles.jpg
https://commons.wikimedia.org/wiki/File:Pyrite-193871_angles.jpg
Photo: Citynoise
Pyritohedron Facet annotated with Face Angles
4 May 2020

14.5
Wikipedia contributors. (2023, January 4).
Dodecahedron.
In *Wikipedia, The Free Encyclopedia*.
Retrieved 11:36, February 7, 2023, from
https://en.wikipedia.org/w/index.php?title=Dodecahedron&oldid=1131592322

14.6
Review
"Quasicrystals and Other Aperiodic Structures
in Mineralogy"
Carlos M. Pina and Victoria López-Acevedo
MDPI Crystals
© 2016 by the authors; licensee MDPI, Basel, Switzerland. This article is an open access article distributed under the terms and conditions of the

Creative Commons Attribution (CC-BY) license
(http://creativecommons.org/licenses/by/4.0/).
16pp

14.7
Wikipedia contributors. (2022, September 14).
Miller index.
In *Wikipedia, The Free Encyclopedia*.
Retrieved 15:07, February 7, 2023, from
https://en.wikipedia.org/w/index.php?title=Miller_index&oldid=1110256
775

14.8
Ho-Mg-Zn dodecahedral quasicrystal, grown by using the self-flux
method (excess Mg), and slowly cooling from 700 C to 480 C. The R-Mg-
Zn family is the first rare-earth containing quasicrystal structure, which
allows the study of localized magnetic moments in a quasiperiodic
environment.
Photograph of a single-grain icosahedral Ho-Mg-Zn quasicrystal grown
from the ternary melt. Shown over a mm scale, the edges are 2.2 mm long.
Note the clearly defined pentagonal facets, and the dodecahedral
morphology. [in: Phys. Rev. B 59, 308–321 (1999)]
By AMES lab., US Department of Energy –
http://cmp.ameslab.gov/personnel/canfield/photos.html#, Public Domain,
https://commons.wikimedia.org/w/index.php?curid=10094632

14.9
"Collisions in outer space produced an icosahedral phase in the Khatyrka
meteorite never observed previously in the laboratory"
Luca Bindi, Chaney Lin, Chi Ma & Paul J. Steinhardt
Scientific Reports volume 6, Article number: 38117 (2016)
https://www.nature.com/articles/srep38117
https://doi.org/10.1038/srep38117

14.10
Wikipedia contributors. (2022, February 20).
Khatyrka meteorite.
In *Wikipedia, The Free Encyclopedia*.
Retrieved 10:02, February 8, 2023, from

https://en.wikipedia.org/w/index.php?title=Khatyrka_meteorite&oldid=1072956653

14.11
File:Zeroth stellation of icosahedron.svg
Tomruen using the Bulatov applet
https://commons.wikimedia.org/wiki/File:Zeroth_stellation_of_icosahedron.png

14.12
Wikipedia contributors. (2023, January 30).
Regular icosahedron.
In Wikipedia, The Free Encyclopedia.
Retrieved 10:35, February 8, 2023, from
https://en.wikipedia.org/w/index.php?title=Regular_icosahedron&oldid=1136440084

CHAPTER FIFTEEN

15.1
"The Curves of Life"
Theodore Andrea Cook MA FSA
Constable of London   1914
Dover of New York   1979
ISBN 0-486-23701-X
512pp

15.2
"The Canon of Polykleitos"
Creative Commons License
The Higher Inquiètude by CG Hughes is licensed under a Creative Commons
Attribution-NonCommercial-ShareAlike 4.0 International License.
Based on a work at higherinquietude.wordpress.com
The Higher Inquiètude
Art History and Visual Culture Blog
https://higherinquietude.wordpress.com/2014/07/04/the-canon-of-polykleitos/

https://blog.artsper.com/en/a-closer-look/ancient-greek-canon-of-proportions/

CHAPTER SIXTEEN

16.1
"Research on mechanical properties and acoustic emission characteristics of rock beams with different lithologies and thicknesses"
Tong-bin Zhao, Peng-fei Zhang , Wei-yao Guoa, Ya-xun Xiao, Yong-qiang Zhao, Xiang Sun

Received August 22, 2021. In revised form October 24, 2021. Accepted October 25, 2021. Available online October 26, 2021. https://doi.org/10.1590/1679-78256711
Latin American Journal of Solids and Structures. ISSN 1679-7825.
Latin American Journal of Solids and Structures, 2021, 18(8), e414 1/16
https://www.scielo.br/j/lajss/a/smfgQXDVxrn8kVmNJGCVc6w/?format=pdf&lang=en

16.2
Wikipedia contributors. (2023, January 22).
Flexural strength.
In Wikipedia, The Free Encyclopedia.
Retrieved 13:13, February 15, 2023, from
https://en.wikipedia.org/w/index.php?title=Flexural_strength&oldid=1135147748

16.3
Wikipedia contributors. (2023, February 11).
Euler's critical load.
In Wikipedia, The Free Encyclopedia. Retrieved 13:34, February 15, 2023, from
https://en.wikipedia.org/w/index.php?title=Euler%27s_critical_load&oldid=1138739267

16.4

<u>Metron Ariston</u> "Metron Ariston" is a book written in 2002 and published on 11 February 2003 in Athens,Greece (275 pages - in Greek). ISBN 960-8286-06-9 Contents: Introduction (about the new measurements made in 2002 and the Megalithic Cubit), Tiryns, the Pyramid of Proetus, Mycenae, Orchomenos (Boiotian), Hyle (Gla), Pylos and Crete, Athens, Parthenon and Stonehenge.Wednesday, January 12, 2011PARTHENON
http://athang1504.blogspot.com/2011/01/parthenon.html
Athang 1504

Athanasios G. Angelopoulos (Username: Athang1504)- I was born in Athens, Greece, on August 19, 1953 and I have studied mathematics, astronomy and ancient Greek language and history. I'm also an amateur painter, sculptor and photographer. In general, I like science, arts, music, sports and everything that is creative. I have spent the last thirty years of my life in research and writing. I have written and published (not self-published) 14 books and many articles that have been read by more than 100,000 people in my country since 1986 (in Greek). They contain only original work based on this research. I believe that I have found some interesting things and, obviously, this is the reason I'm presenting some of them here (in English).

16.5
"On the Solution to Octic Equations"
Raghavendra G. Kulkarni
Follow this and additional works at: https://scholarworks.umt.edu/tme
Part of the Mathematics Commons
Kulkarni, Raghavendra G. (2007) "On the Solution to Octic Equations,"
The Mathematics Enthusiast: Vol. 4: No. 2 , Article 6.
DOI: https://doi.org/10.54870/1551-3440.1071
Available at: https://scholarworks.umt.edu/tme/vol4/iss2/6

CHAPTER SEVENTEEN

17.1
"The Pi-Phi Product"
Ed Oberg and Jay A Johnson
7 April 2000
An eleven-page descriptive .doc paper
and a 33-iteration .xls worksheet are available to
download at
https://www.goldennumber.net/pi-phi-fibonacci/

CHAPTER EIGHTEEN

18.1
Wikipedia contributors. (2023, July 9).
Binomial coefficient.
In Wikipedia, The Free Encyclopedia.
Retrieved 09:58, July 9, 2023, from
https://en.wikipedia.org/w/index.php?title=Binomial_coefficient
&oldid=1164408748

18.2
Wikipedia contributors. (2023, June 20).
Bernoulli number.
In Wikipedia, The Free Encyclopedia.
Retrieved 08:22, June 28, 2023, from
https://en.wikipedia.org/w/index.php?title=Bernoulli_number
&oldid=1161052336

Table 23.2 p810 Bernoulli and Euler Numbers
"Handbook of Mathematical Functions With Formulas, Graphs and
 Mathematical Tables"
Edited by Milton Abramowitz and Irene A Stegun
National Bureau of Standards
Applied Mathematics Series . 55
Washington, DC
Issued June 1964: Tenth Printing, December 1972, with corrections
pp 1060

18.3
Plouffe Formula for PI
"Formula for the n'th digit of π and π^n"
Simon Plouffe
January 29, 2022
7pp

CHAPTER NINETEEN

19.1
John 8:32

APPENDICES

APPENDIX ONE
SOME CONVENIENT SURDS OF TRIGONOMETRIC FUNCTIONS

LHS	RHS
$\sin\left(\dfrac{1\cdot\pi}{10}\right)=0.309$	$\dfrac{\sqrt{5}-1}{4}=0.309$
$\sin\left(\dfrac{2\cdot\pi}{10}\right)=0.588$	$\dfrac{\sqrt{10-2\sqrt{5}}}{4}=0.588$
$\sin\left(\dfrac{3\cdot\pi}{10}\right)=0.809$	$\dfrac{1+\sqrt{5}}{4}=0.809$
$\sin\left(\dfrac{4\cdot\pi}{10}\right)=0.951$	$\dfrac{\sqrt{10+2\sqrt{5}}}{4}=0.951$
$\sin\left(\dfrac{5\cdot\pi}{10}\right)=1$	1
$\sin\left(\dfrac{6\cdot\pi}{10}\right)=0.951$	$\dfrac{\sqrt{10+2\sqrt{5}}}{4}=0.951$
$\cos\left(\dfrac{1\cdot\pi}{10}\right)=0.951$	$\dfrac{\sqrt{10+2\sqrt{5}}}{4}=0.951$
$\cos\left(\dfrac{2\cdot\pi}{5}\right)=0.309$	$\dfrac{\sqrt{5}-1}{4}=0.309$
$\cos\left(\dfrac{3\cdot\pi}{5}\right)=-0.309$	$\dfrac{1-\sqrt{5}}{4}=-0.309$
$\cos\left(\dfrac{4\cdot\pi}{5}\right)=-0.809$	$-\left(\dfrac{1+\sqrt{5}}{4}\right)=-0.809$
$\cos\left(\dfrac{4\cdot\pi}{10}\right)=0.309$	$\dfrac{\sqrt{5}-1}{4}=0.309$
$\cos\left(\dfrac{5\cdot\pi}{10}\right)=6.123\cdot10^{-17}$	0
$\cos\left(\dfrac{6\cdot\pi}{10}\right)=-0.309$	$\dfrac{1-\sqrt{5}}{4}=-0.309$

SELECTED ALGEBRAIC POLYNOMIALS IN THE MINOR RATIO OF PHIDIAS, ϕ

9	$4 \cdot \phi^4 + 6 \cdot \phi^3 + 4 \cdot \phi^2 + 4 \cdot \phi + 3 = 9$
9	$2 \cdot \phi^4 + 3 \cdot \phi^3 + 8 \cdot \phi^2 + 8 \cdot \phi = 9$
9	$4 \cdot \phi^2 + 4 \cdot \phi + 5 = 9$
$\phi^3 + 7 = 7.23606797749979$	$2 \cdot (\phi + 1)^2 \cdot (\phi^2 + 1) = 7.23606797749979$
$\phi^3 + 7 = 7.23606797749979$	$2 \cdot \phi^4 + 4 \cdot \phi^3 + 4 \cdot \phi^2 + 4 \cdot \phi + 2 = 7.23606797749979$
7	$2 \cdot \phi^4 + 3 \cdot \phi^3 + 4 \cdot \phi^2 + 4 \cdot \phi + 2 = 7$
5	$2 \cdot \phi^4 + 3 \cdot \phi^3 + 4 \cdot \phi^2 + 4 \cdot \phi = 5$
5	$4 \cdot \phi^2 + 4 \cdot \phi + 1 = 5$
4	$4 \cdot \phi^2 + 4 \cdot \phi = 4$
4	$8 \cdot \phi^4 + 12 \cdot \phi^3 = 4$
$\sqrt{\phi^3 + 7} = 2.68999404785583$	$\sqrt{2} \cdot (\phi + 1) \cdot \sqrt{\phi^2 + 1} = 2.68999404785583$
	$2 \cdot \phi^4 + 3 \cdot \phi^3 + 4 \cdot \phi^2 = 2.52786404500042$
3	$2 \cdot \phi^4 + 3 \cdot \phi^3 + 4 \cdot \phi^2 + 4 \cdot \phi - 2 = 3$
3	$4 \cdot \phi^4 + 6 \cdot \phi^3 + 1 = 3$
3	$6 \cdot \phi^4 + 9 \cdot \phi^3 = 3$
2	$2 \cdot \phi^4 + 3 \cdot \phi^3 + 1 = 2$
2	$4 \cdot \phi^4 + 6 \cdot \phi^3 = 2$
2	$2 \cdot \phi^3 \ (2 \cdot \phi + 3) = 2$
1	$2 \cdot \phi^3 \cdot \left(\phi + \dfrac{3}{2}\right) = 1$
1	$2 \cdot \phi^4 + 3 \cdot \phi^3 = 1$
1	$\phi^2 + \phi = 1$

APPENDIX THREE
SELECTED POWERS OF ONE HALF OF THE MAJOR RATIO OF PHIDIAS, Φ/2

Φ	1.618033988749890	
z = Φ/2	0.809016994374947	

Power p	z^p
0	1.000000000000000
1	0.809016994374947
2	0.654508497187474
3	0.529508497187474
4	0.428381372890605
5	0.346567810742171
6	0.280379248593737
7	0.226831576982411
8	0.183510600639640
9	0.148463194565423
10	0.120109247442621
11	0.097170422362666
12	0.078612523041989
13	0.063598867111661
14	0.051452564316328
15	0.041625998936079
16	0.033676140547121
17	0.027244570007580
18	0.022041320140571
19	0.017831802572180
20	0.014426231321233
21	0.011671066303662
22	0.009442090982139
23	0.007638812066985
24	0.006179928779027
25	0.004999667406260
26	0.004044815897887
27	0.003272324800508
28	0.002647366374726
29	0.002141764387490
30	0.001732723787426
31	0.001401802990586
32	0.001134082442149

TABLE OF FIBONACCI NUMBERS ON THE RATIOS OF PHIDIAS

| φ | 0.618033988749895 |
| Φ | 1.618033988749890 |

n	Fφ (n)	FΦ (n)
1	2.618033988749890	1.618033988749890
2	1.618033988749890	2.618033988749890
3	1.000000000000000	4.236067977499790
4	0.618033988749895	6.854101966249680
5	0.381966011250105	11.090169943749500
6	0.236067977499790	17.944271909999200
7	0.145898033750316	29.034441853748600
8	0.090169943749474	46.978713763747800
9	0.055728090000841	76.013155617496400
10	0.034441853748633	122.991869381244000
11	0.021286236252208	199.005024998741000
12	0.013155617496425	321.996894379985000
13	0.008130618755783	521.001919378725000
14	0.005024998740641	842.998813758710000
15	0.003105620015142	1364.000733137440000
16	0.001919378725500	2206.999546896150000
17	0.001186241289642	3571.000280033580000
18	0.000733137435857	5777.999826929730000
19	0.000453103853785	9349.000106963310000
20	0.000280033582073	15126.999933893000000
21	0.000173070271712	24476.000040856400000
22	0.000106963310360	39602.999974749400000
23	0.000066106961352	64079.000015605700000
24	0.000040856349008	103681.999990355000000
25	0.000025250612343	167761.000005961000000
26	0.000015605736665	271442.999996316000000
27	0.000009644875678	439204.000002277000000
28	0.000005960860987	710646.999998593000000
29	0.000003684014692	1149851.000000870000000
30	0.000002276846295	1860497.999999460000000
31	0.000001407168397	3010349.000000330000000
32	0.000000869677897	4870846.999999790000000

Reference:
Nizhizawa: "Refined Quadratic Estimators of Shafer's Inequality" 2017

The Nizhizawa Process

Allow that:-

$$z = \frac{\Phi}{2}$$

Equation E.1

then:-

$$AT_{fido} = \tan^{-1} z$$

Equation E.2

Now:-

$$t = \sqrt{\frac{1 - z^2}{z}}$$

Equation E.3

whilst:-

$$v = \sqrt{(\pi^2 - 4)^2 + (2\pi t)^2}$$

Equation E.4

and:-

$$w = \sqrt{32 + (2\pi t)^2}$$

Equation E.5

Hence the Lower Bound, t_{LB}, and the Upper Bound, t_{UB}, are given by:-

$$t_{LB} = \frac{\pi^2 z}{4 + v}$$

Equation E.6a

$$t_{UB} = \frac{\pi^2 z}{4 + w}$$

Equation E.6b

Then the Arithmetic Mean, am, of t_{LB} and t_{UB} estimates the inverse tangent of $z = \Phi/2$.

PSD(AT$_{fido}$,am) is -3.40589887385864

The Qiaol-Chen Process

Reference:
Quan-Xi Qiaol and Chao-Ping Chen "Approximations to Inverse Tangent Function"

This boundary estimator due to Quan-Xi Qiaol and Chao-Ping Chen reports that:-

$$t_{LB} = z\left[\frac{\pi}{2} - \tan^{-1}(z)\right]$$

Equation E.7a

$$t_{UB} = \frac{z^2 + \frac{4}{15}}{z^2 + \frac{3}{5}}$$

Equation E.7b

Self-evidently, the method is predicated upon knowledge of Tan^{-1}(z), the very object of our determination.

I calculated the Arithmetic Mean, μ_{QC}, and the Geometric Mean, g_{QC}, as:-

$$\mu_{QC} = \frac{t_{LB} + t_{UB}}{2}$$

and:-

$$g_{QC} = \sqrt{t_{LB} \cdot t_{UB}}$$

The Qiaol-Chen bounds of Tan-1(z) were:-

$\text{PSD}(\text{AT}_{fido}, \mu_{QC}) = -6.93583542474449$
$\text{PSD}(\text{AT}_{fido}, g_{QC}) = -6.93102674591988$

INDEX

PIPHI

WARREN